International Conference on Advanced Materials and Modern Manufacturing (ICAMMM 2021)

AA007593

IOP Conference Series: Materials Science and Engineering
Volume 1183

Chennai, India
26 – 28 March 2021

Editors:

A. Devaraju
D. Palanisamy

ISBN: 978-1-7138-4643-7
ISSN: 1757-8981

Printed from e-media with permission by:

Curran Associates, Inc.
57 Morehouse Lane
Red Hook, NY 12571

Some format issues inherent in the e-media version may also appear in this print version.

This work is licensed under a Creative Commons Attribution 3.0 International Licence.
Licence details: http://creativecommons.org/licenses/by/3.0/.

No changes have been made to the content of these proceedings. There may be changes to pagination and minor adjustments for aesthetics.

Printed with permission by Curran Associates, Inc. (2022)

For permission requests, please contact the Institute of Physics at the address below.

Institute of Physics
Dirac House, Temple Back
Bristol BS1 6BE UK

Phone: 44 1 17 929 7481
Fax: 44 1 17 920 0979

techtracking@iop.org

Additional copies of this publication are available from:

Curran Associates, Inc.
57 Morehouse Lane
Red Hook, NY 12571 USA
Phone: 845-758-0400
Fax: 845-758-2633
Email: curran@proceedings.com
Web: www.proceedings.com

TABLE OF CONTENTS

IOP Conf. Series: Materials Science and Engineering 1183 (2021) 011001 doi:10.1088/1757-899X/1183/1/011001

Dr.A. Devaraju,
Dr.D. Palanisamy, Editor(s)

International Conference on Advanced Materials & Modern Manufacturing

ICAMMM 2021

Content from this work may be used under the terms of the Creative Commons Attribution 3.0 licence. Any further distribution of this work must maintain attribution to the author(s) and the title of the work, journal citation and DOI.
Published under licence by IOP Publishing Ltd

IOP Conf. Series: Materials Science and Engineering 1183 (2021) 011001 doi:10.1088/1757-899X/1183/1/011001

PREFACE

We are pleased to bring before you the proceedings of the International Conference on Advanced Materials and Modern Manufacturing (ICAMMM 2021) which was scheduled during 26th to 28th March 2021.

The ICAMMM 2021 is organized by the Department of Mechanical Engineering, Adhi college of Engineering and Technology, Tamil Nadu, India.

The conference aims to bring together the researchers, faculty, students and industrial professionals from all over the world to share their research findings, experience and to discuss the emerging issues facing in different spheres of Mechanical Engineering.

This conference focuses on today's technical challenges, research updates and breakthrough innovations that are shaping the future of Mechanical Engineering. ICAMMM 2021 will provide a platform for sharing and exchange of ideas and information between all the stake holders of technical education and research.

The selected and peer reviewed papers of ICAMMM 2021 focused on Advanced Materials, Modern Manufacturing, Thermal Engineering, Industrial Engineering, Tribology, Sustainable Energy, vibration and Control etc…

I would like to thank the organizing committee members, Co-ordinators, staff, students and reviewers for their continuous support and guidance without whom; all this would not have been possible.

I hope that all researchers will benefit from the proceedings of ICAMMM 2021, which will enrich their knowledge with the advanced technology developments.

IOP Conf. Series: Materials Science and Engineering 1183 (2021) 011001 doi:10.1088/1757-899X/1183/1/011001

Patron(s)
Dr. S. Dilli Saran raj, Chairman
Smt. Saranya Saran, Correspondent

Co-Patron
Smt. Sujatha Maran, CEO

Convener(s)
Dr. A. Devaraju, Principal
Dr. D. Palanisamy, Associate Professor/ Head

Organizing Committee
Dr. V. Paranthaman, Associate Professor/Vice Principal
Mr. P. Murali, Asst. Professor
Mr. D. Umapathi, Asst. Professor
Mr. R. Harikumar, Asst. Professor
Mr. R. Prabu, Asst. Professor
Mr. R. Gopi, Asst. Professor
Mr. A. Raji, Asst. Professor
Mr. D. Gopinath, Asst. Professor
Mr. P. Kumaravel, Asst. Professor
Mr. A. Fantin Arokiaraj, Asst. Professor
Mr. S. Sivakumar, Asst. Professor
Mr. A. Suresh, Asst. Professor

National Technical Advisory Committee
Dr. M. Kamaraj, IITM, Chennai
Dr. P. Sathya, NIT, Tiruchirapalli
Dr. K. V. N. Surendra, IIT, Palakkad
Dr. B. S. S. Daniel, IIT, Roorkee
Dr. S. Basavarajappa, IIIT, Dharwad
Dr. P. Senthil, NIT, Tiruchirapalli
Dr. C. Sathya Narayanan, NIT, Tiruchirapalli
Dr. K. Noel Jacob, NIT, Calicut
Dr. P. S. Balaji, NIT. Rourkela
Dr. K.E.K. Vimal, NIT, Patna
Dr. A. Karpagaraj, NIT, Patna

Dr. C. Velmurugan, IIIT, Tiruchirapalli
Dr. S. C. Vettivel, CCET, Chandigarh
Dr. V. R. Rajeev, CET, Trivandrum
Dr. P. C. Krishnamachary, S.V.E.C, (Autonomous) Tirupati
Dr. S. Aravind Raj, VIT, Vellore
Dr. B. Vijaya, Dr. MGR Research Institute, Chennai
Dr. Shashi Bahl, IK Gujral Technical University, Punjab
Dr. R. Gopinath, SR University, Warangal
Dr. A. Elayaperumal, Anna University, Chennai
Dr. V. Balasubramanian, Annamalai Univ, Chidambaram
Dr. M. Manohar, ISRO, Trivandrum
Dr. K. C. Varaprasad, S.V.E.C,(Autonomous) Tirupati
Dr. N. Manikandan, S.V.E.C,(Autonomous) Tirupati
Dr. J.S. Binoj, S.V.E.C,(Autonomous) Tirupati
Dr. M. S. Alphin, SSNCE, Chennai
Dr. Ramesh raju, SREC, Nandyal
Dr. M. Natraj, GCE, Tirunelveli
Dr. R. Edwin Raj, SXCCE, Nagercoil
Dr. A. Sagai Francis Britto, RCET, Anjugramam
Dr. Jayakrishnan Kanthasamy, VIT, Vellore
Dr. Pankaj N Dhatrak, MIT-WPU, Pune
Dr. P. Murthi, SR University, Warangal, Telangana
Dr. C.R. Raghavendra, Govt. Engg. College, Karnataka
Dr. D. Jafrey Daniel James, KRCE, Trichy
Dr. K. Lenin, KRCE, Trichy
Dr. N.L. Parthasarathi, IGCAR, Kalpakkam
Mr. Gopinath Chakravarthy, Rolls Royce India (P) Ltd, Bangalore

International Advisory Committee

Dr. Abu Bakar Mahat, Infrastructure University, Malaysia
Dr. Manoj Gupta, National University of Singapore (NUS)
Dr. Parvathy Rajendran, Universiti Sains, Malaysia
Dr. S. Sivasankaran, Qassim University, Saudi Arabia
Dr. Mohammad Nazri Mohd Jaafar, Universiti Teknologi, Malaysia
Dr. M. S. J. Hashmi, Dublin City University, Ireland
Dr. Isiaka O. Bakare, Rubber Research Instititute, Nigeria
Dr. Sandro Campos Amico, UFRGS, Brazil

IOP Conf. Series: Materials Science and Engineering 1183 (2021) 011001 doi:10.1088/1757-899X/1183/1/011001

Dr. Mohd Sapuan Salit, Universiti Putra, Malaysia
Dr. Sukur Bin Abu Hasan, Universiti teknologi, Malaysia
Prof. Dr. M. Mariatti Jaafar, University Of Science, Malaysia
Dr. Ing. Habil. Sucharat Siengchin, Bangkok, Thailand
Dr. M. R. Sanjay, KMUTNB, Bangkok, Thailand
Dr.Ian J Davies, KMUTNB, Bangkok, Thailand
Dr. Dinganta B Das, Loughborough University
Dr. Cheng Guiping, Nilai University, Malaysia
Dr. Ajit Pal Singh, Defense University, Ethiopia
Dr. Prabhu Paramasivam, Mettu University, Ethiopia
Dr. Ranjeet Rajan, Ruwais Men's College, UAE
Dr. S. Porchilamban, Nizwa College Of Technology, Oman
Prof. Siva Marimuthu, Nilai University, Malaysia

List of Reviewers
Dr. K.E.K. Vimal, NIT, Patna
Dr. S. Aravind Raj, VIT, Vellore
Dr. B. Vijaya, Dr. MGR Research Institute, Chennai
Dr. Shashi Bahl, IK Gujral Technical University, Punjab
Dr. R. Gopinath, SR University, Warangal
Dr. J.S. Binoj, S.V.E.C,(Autonomous) Tirupati
Dr. D. Jafrey Daniel James, KRCE, Trichy
Dr. K. Lenin, KRCE, Trichy
Dr. D. Arul Kirubakaran, Karunya Institute of Technology and Science, Coimbatore
Dr. R. Pugazhenthi, VISTAS, Chennai

IOP Conf. Series: Materials Science and Engineering 1183 (2021) 011002 doi:10.1088/1757-899X/1183/1/011002

Peer Review Declaration

All papers published in this volume of *IOP Conference Series: Materials Science and Engineering* have been peer reviewed by the expert committee members administered by the editiors. Reviews were conducted with referees to the professional and scientific standards expected of a proceedings journal published by IOP Publishing.

Manuscripts are selected based on the reviewer's comments on the criteria given to them as.

- Accept as it is
- Accept with minor changes
- Accept with revision as noted in comments
- Reject in current form, but may be resubmitted
- Reject, with no resubmission

Authors are given opportunity to revise the manuscript based on the comments they received. Selection is made on the step by step manner as mentioned below

Step 1: Initial Proof read by conference coordinators for scope

Step 2: Plagiarism check using "Turnitin" Software

Step 3: Paper sent to reviewers.

Step 4: Send back the review commands to authors for revision.

Step 5: Send back the revised manuscript to the reviewers for their approval

Step 6: Finalize the paper for proceeding

- Type of Peer Review **Double-blind**

Content from this work may be used under the terms of the Creative Commons Attribution 3.0 licence. Any further distribution of this work must maintain attribution to the author(s) and the title of the work, journal citation and DOI.
Published under licence by IOP Publishing Ltd

IOP Conf. Series: Materials Science and Engineering 1183 (2021) 011002 doi:10.1088/1757-899X/1183/1/011002

• Conference Submission management system	**Author ↔Editor ↔Reviewer** **E-mail Submission**
• Number of Submissions received	**17**
• Number of Submissions sent for review	**13**
• Number of Submissions accepted	**08**
• Acceptance Rate	**47%**
• Average number of review per paper	**2**
• Total number of reviewers involved	**10**
• Any additional info on review process	**All the manuscripts received are checked for plagiarism using "Turnitin Software".**
• Contact person for queries	**Dr. D. Palanisamy** **icammm@adhi.edu.in**

IOP Conf. Series: Materials Science and Engineering 1183 (2021) 012001 doi:10.1088/1757-899X/1183/1/012001

Optimization for Injection Moulding process parameters towards Warpage and Shrinkage of HDPE-PBI composites

M. Ramesh[1,2], **K. Panneerselvam**[2*]

[1]Department of Mechanical Engineering, K. Ramakrishnan College of Engineering, Trichy.
[2]Department of Production Engineering, NIT, Trichy.

[*]Corresponding author Email: kps@nitt.edu

Abstract. Injection moulding is a high-quality process that is preferred in modern manufacturing. Due to the defects in HDPE polymer composites the parameter's had to be adjusted based on the necessity. Therefore, this study of parameters is to improve the quality characteristics of the injection moulding of HDPE-PBI composite samples. As the composition of HDPE composites varies, the quality aspects that must be controlled are shrinkage and warpage. The parameters that are considered as factors are Melting Temperature (°C), Injection pressure (MPa) and cooling time (sec). Using the Taguchi Optimization method of L9 Orthogonal array, the injection moulding process was performed and found that the injection pressure of 65 MPa and temperature of 180-185 °C showed better results. The influence of parameters was studied through this research using Analysis of Variance (ANNOVA).

Keywords: *HDPE Composite, Injection Moulding Machine, Optimization, ANNOVA*

1. Introduction

The automotive industry is working to improve gas mileage by replacing some metal parts with polymer composites. The injection moulding process has grown in popularity in the polymer processing industry. This method is used for mass production at a low cost and time [1]. The plastic HDPE-PBI composite pellets are put in the hopper of an injection moulding machine. The pellets are heated by four heating zone coils. A screw is provided inside the barrel to provide injection pressure and melting of plastics forced into the mould cavity. Warpage and shrinkage must be kept to a minimum because they affect the product's quality. To find the optimum condition of processing of the HDPE composites, to achieve desired properties, optimized processing conditions are necessary for the Injection moulding machines [2]. The warpage is caused by the shrinkage variation of plastic parts by the injection process. It is found that area shrinkage (due to packing pressure and crystallization variation), differential cooling (caused due to sectional shrinkage) and flow orientation (shrinkage caused due to the direction of plastic injection) [3-5]. The various factors that control these characteristics are packing pressure, injection pressure, melt temperature, screw speed, mould temperature, holding time and cooling time. The warpage problem was discussed by Ming-chih Huang et.al (1999) with the factors that affected the moulded parts. S.H. tang et al. (2006) have found that warpage is caused by some important parameters, like packing time and melt temperature, along with filling time and packing pressure. The method that is used for conducting multiple experiments is the Design of Experiments. DoE is used to reduce the number of experiments to obtain the optimal

Content from this work may be used under the terms of the Creative Commons Attribution 3.0 licence. Any further distribution of this work must maintain attribution to the author(s) and the title of the work, journal citation and DOI.
Published under licence by IOP Publishing Ltd

IOP Conf. Series: Materials Science and Engineering 1183 (2021) 012001 doi:10.1088/1757-899X/1183/1/012001

parameters [6]. The Taguchi technique is a design of experiment (DOE) that uses a statistical method to optimize process parameters while minimizing variability and improving product quality. Nominal is the best of the Taguchi techniques in this study and is an appropriate application based on data characterization. It explains the techniques and steps used in DOE to determine the best quality parameter for a given quality characterization [7].

The major plastic material in use is the commodity polymer HDPE. From literature, we have found that the main reason for defects in the quality of plastic products is the setting up of process parameters in the injection moulding machine. This research investigation is about finding the influencing factors affecting shrinkage and warpage of HDPE composites.

2. Injection moulding machine

An injection moulding machine is one of the processes of melting plastic pellets and creating products of high quality and efficiency. The machine consists of four zones. Each zone is made up of heating coils that vary the temperature based on the plasticization. The front zone consists of a suck-back section, which is the nozzle of the machine. It is situated at the forefront of the barrel. There will be some internal pressure in the nozzle. The temperature had to be lower as compared to that of the plasticization area. This temperature setting is lower and not too low because it is to stop the flow of molten plastic under internal pressure [8]. Additional plasticization pressure is required to compensate for the flow of water back if the temperature is kept too low. The centre zone is the barrel middle zone. Here the temperature has to rise above the melting point of plastic and its sole purpose is to make the plastic into a molten state. The noticeable thing is the temperature should be too high as the decomposition of plastic material might happen. If the temperature is low in this zone, the plasticization process will not be proper and the torque has to be increased for the screw. The lower temperature zone is the rear zone, which is the feeding area near the hopper. The temperature set should be less than that of the centre zone, i.e., less than the melting temperature of the plastic pellets. This is also known as the pre-heating zone, where the temperature begins to rise towards the melting point. If the temperature is above the melting point, then the plastic starts to get melted and the process gets altered as the screw slippage occurs and, in the end, affects the quality of the material and the material delivered. If the pre-heating temperature is too low, there should be an increase in screw torque.

3. Data collection

The data for these studies is to calculate the existing mould dimensions and sample dimensions using a digital vernier caliper and a dial gauge. The shrinkage and warpage in an injection moulding process determine the quality of the product output.The specimens that are removed from the mould after the injection moulding process is completed must be cleared of flash before being measured.

The shrinkage can be calculated from equation 1.

$$S = \frac{Ma-Mp}{Ma} \text{ ----------------------- } (1)$$

The Warpage can be calculated from equation 2.

$$W = h - t_p \text{ -----------------------} (2)$$

S – Shrinkage value (mm)
M_a- Mould dimension (mm)
M_p – Specimen dimension taken from mould (mm)
W – Warpage value (mm)
h- Maximum height of plate bend (mm)
t_p – thickness of plate (mm)

4. Parameters selected and Experimental design

The taguchi method was opted to do this design of experiments; the three levels of parameters were selected for factors like Injection pressure, injection temperature, Packing pressure and cooling time.

The three values shows the low, medium and high values which is significant factors in increasing the quality characteristics of the output product. The table shows the factors and levels of parameter taken for this research. The L9 Orthogonal array had been adopted for this work as it is having three levels and three factors. The table shows the L9 Orthogonal array, whereby A is for Injection pressure, B is for melting temperature, C is for cooling time [9-11].

Table 1. Process parameters and levels

No	Factors	Level 1	Level 2	Level 3
1	Injection Pressure – A (MPa)	60	65	70
2	Melting Temperature – B (°C)	180	190	200
3	Cooling time – C (sec)	10	15	20

Table 2. Warpage and Shrinkage value for HDPE composites by L9 Orthogonal Array

Trial No.	A	B	C	Shrinkage (mm)	Warpage (mm)
1	60	180	10	0.023	0.076
2	60	190	15	0.025	0.079
3	60	200	20	0.029	0.085
4	65	180	15	0.011	0.068
5	65	190	20	0.016	0.074
6	65	200	10	0.018	0.071
7	70	180	20	0.016	0.072
8	70	190	10	0.013	0.07
9	70	200	15	0.026	0.075

Table 3. Analysis of Variance for Shrinkage

Source	DF	Seq SS	Adj MS	F_{cal}	% of contribution
Injection Pressure (MPa)	2	0.000203	0.000101	25.33	74.6
Temperature (C)	2	0.000053	0.000026	6.58	19.4
Cooling Time (s)	2	0.000009	0.000004	1.08	3.33

IOP Conf. Series: Materials Science and Engineering 1183 (2021) 012001 doi:10.1088/1757-899X/1183/1/012001

ERROR	2	0.000008	0.000004	2.94
TOTAL	8	0.000272		100

From the table 3, we can infer the influence of parameters from the percentage of contribution in the shrinkage values. The higher being the injection pressure which contributes about 74.6%. The temperature has a percentage of 19.4 and it's the 2nd influential contributor in the increase of shrinkage value. The cooling time parameter which is considered is not much influential comparatively to temperature and injection pressure. The R-square value obtained for shrinkage is 97.06 % from the ANNOVA.

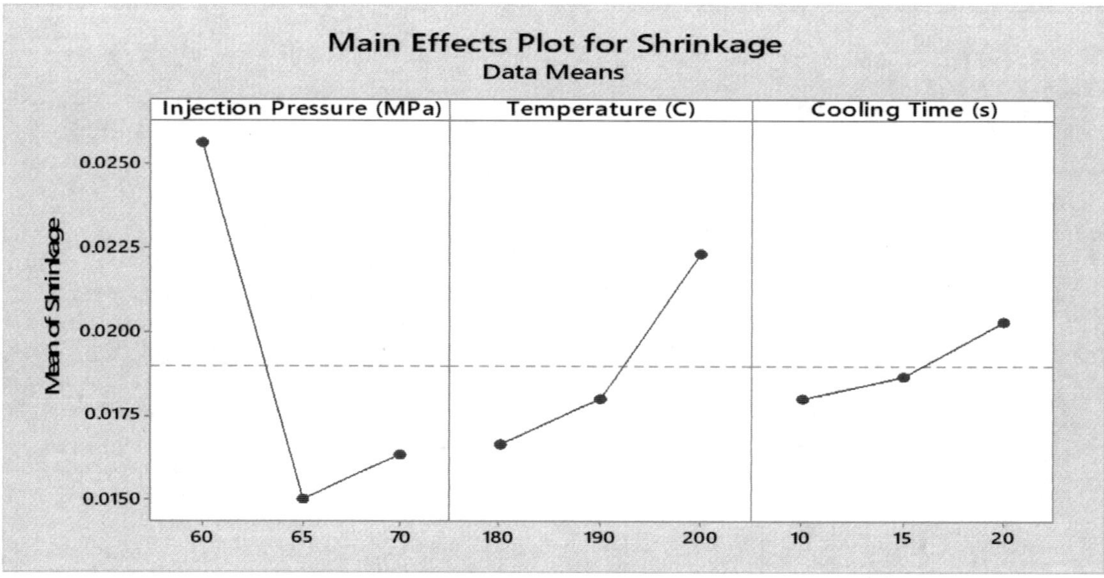

Figure 1. Main Effects plots for Shrinkage

The fig 1 shown is the main effect plots of shrinkage of the HDPE-1% PBI samples. The highest shrinkage was seen at the Injection pressure of 60 MPa with a value of 0.0250 mm. the lowest shrinkage was obtained at the 65 MPa and at 70MPa the shrinkage increased slightly to 0.0163 mm. the decrease of shrinkage is about 45 % when the highest values and lowest values are compared with each other. From the temperature vs Shrinkage graph, the highest value is recorded at 200 °C. The lowest shrinkage was seen at 180°C. The increase in the shrinkage values were noticed as the temperature profile increases. The temperature required being optimum and little above the melting temperature, so that minimum shrinkage can be obtained, if the temperature is too high the shrinkage will be more and also the materials get burned up and degrades. The cooling time didn't have much of an impact when compared to Injection pressure (MPa) and Temperature (°C). There was a little increase while the cooling time was increased from 10 seconds to 20 seconds. This time was given after the machine cooling time, which is 50 seconds. The Packing pressure which was considered in other journals was not taken into account as the packing pressure was determined by some percentage pressure of Injection pressure. So it will directly equal to injection pressure shrinkage values, therefore the packing pressure was not taken into account in this study.

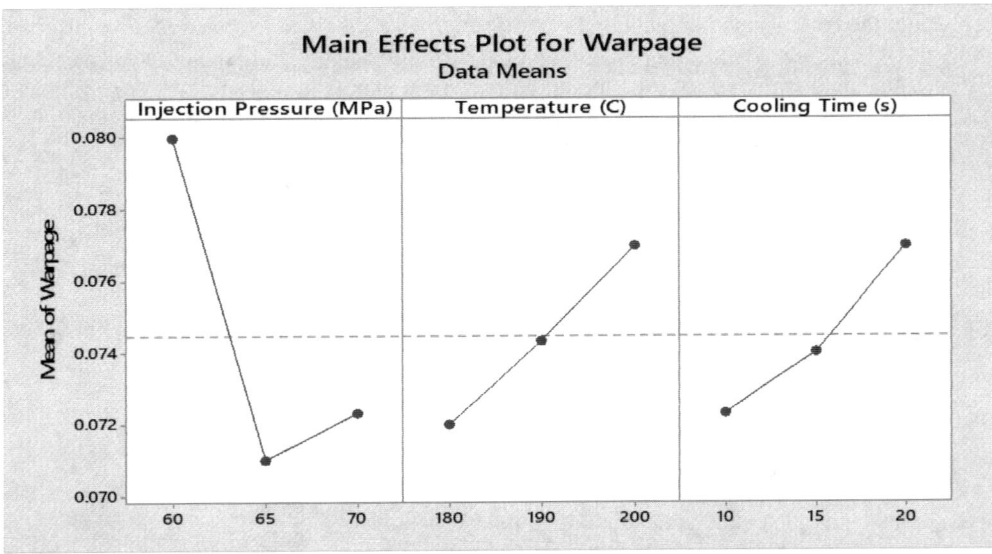

Figure 2. Main Effects plots for Warpage

The fig 2 shown is the main effect plots of warpage of the HDPE-1% PBI samples. The lowest warpage was seen at the Injection pressure of 65 MPa with a value of 0.080 mm. the highest warpage was obtained at the 60 MPa and the warpage increased slightly to 0.072 mm at 70 MPa. The warpage is a defect of bending due to variations in the temperature inside the mould. The side of sample that was near the injection sample is the hot area when compared with the side of the sample that was sticking on the base of die. This difference in the hot and cold region will cause the warpage effect in the injection moulded products [12]. The temperature vs warpage graph clearly states the temperature at which the warpage was high was the highest temperature of study. The graph was linear in nature and stating that, as the temperature increases, the difference of hot and cold region will be more and in turn the warpage defect will also be high. The warpage should be minimum for the proper output of products to be obtained [13].

The cooling time didn't have much of an impact when compared to Injection pressure (MPa) and Temperature (°C). There was a little increase while the cooling time was increased from 10 seconds to 20 seconds. This time was given after the machine cooling time, which is 50 seconds.

Table 4. Analysis of Variance for Warpage

Source	DF	Seq SS	Adj MS	F_{cal}	% of contribution
Injection Pressure (MPa)	2	0.000142	0.000071	91	66.3
Temperature (C)	2	0.000038	0.000019	24.14	17.75
Cooling Time (s)	2	0.000034	0.000017	21.57	15..8
ERROR	2	0.000002	0.000001		0.93
TOTAL	8	0.000214			100

From the table 4, the percentage of contribution of each parameters are shown, we can infer the influence of parameters from the percentage of contribution in the warpage defect. The warpage is due to the temperature difference while injecting the molten plastic into mould. The major contributor to this warpage is the Injection pressure (MPa) [14]. The temperature contributes about 17.75 % and the cooling time contributes 15.8 %. Both these parameters are nearly sharing equal contribution in the influence of Warpage. The R-square value obtained from the ANNOVA table is 99.27%.

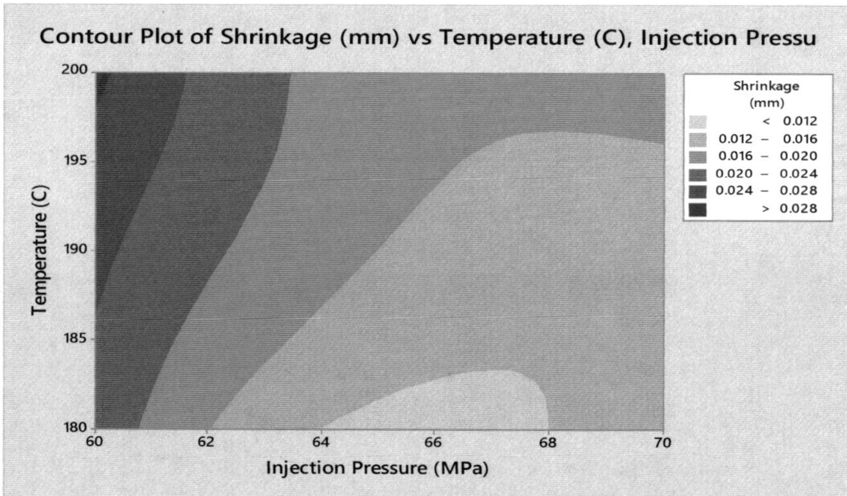

Figure 3. Graph for Shrinkage vs Temperature and Injection pressure for HDPE composite

From the Contour plot graph, the darker regions shows the maximum effect of shrinkage, which is caused due to the reduction in injection pressure and increase in temperature [15]. The less shrinkage effect is seen at Injection pressure of 64-68 MPa and Temperature of 180-183°C.

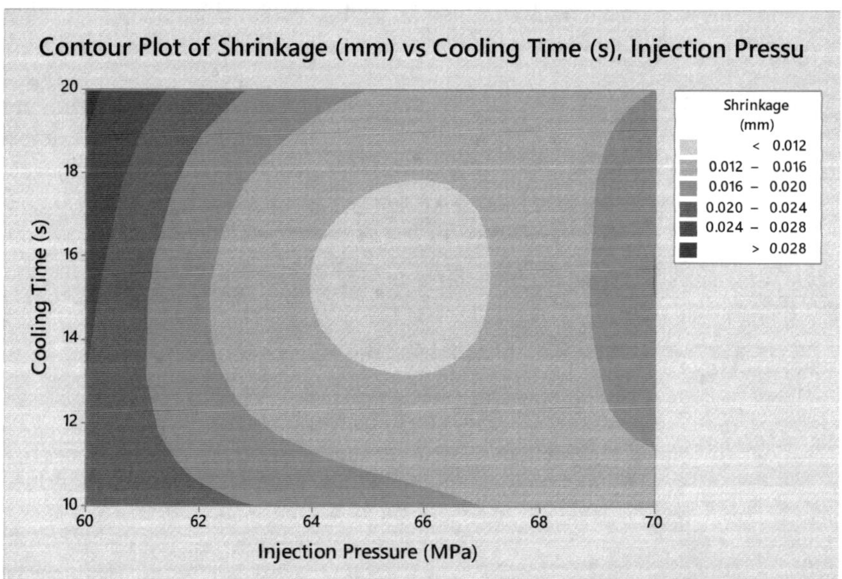

Figure 4. Graph for Shrinkage vs Cooling time and Injection pressure for HDPE composite

IOP Conf. Series: Materials Science and Engineering 1183 (2021) 012001 doi:10.1088/1757-899X/1183/1/012001

From the Contour plot graph 4, the injection pressure of 64-67 MPa with Cooling time of 13-18 seconds had lesser attributes of shrinkage, as the pressure decreases the shrinkage value gets increased.

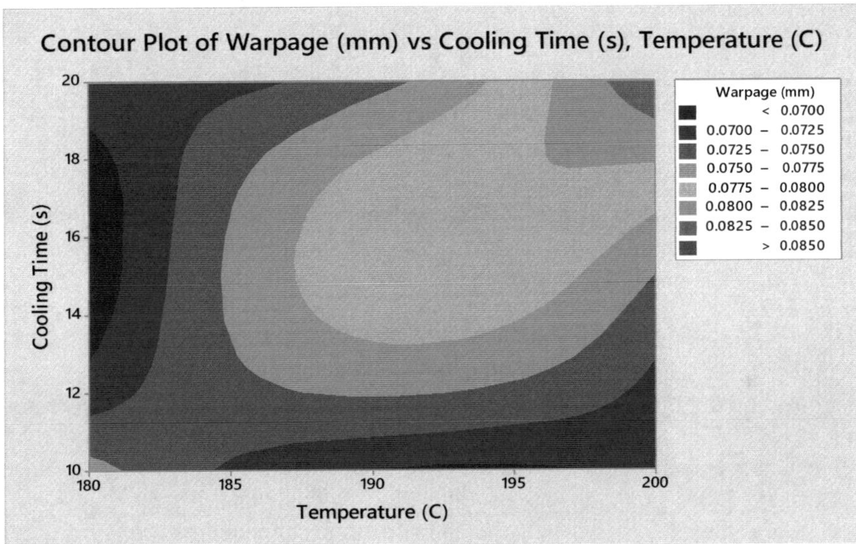

Figure 5. Graph for Warpage vs Temperature and Cooling time for HDPE composite

From the contour plot graph, the warpage is higher at the region of higher temperature and more cooling, time that is during the temperature is above 195-200 °C. The lower the temperature and more the cooling time, the warpage is very less. The 180 C have the betterquality parameter for reducing warpage.

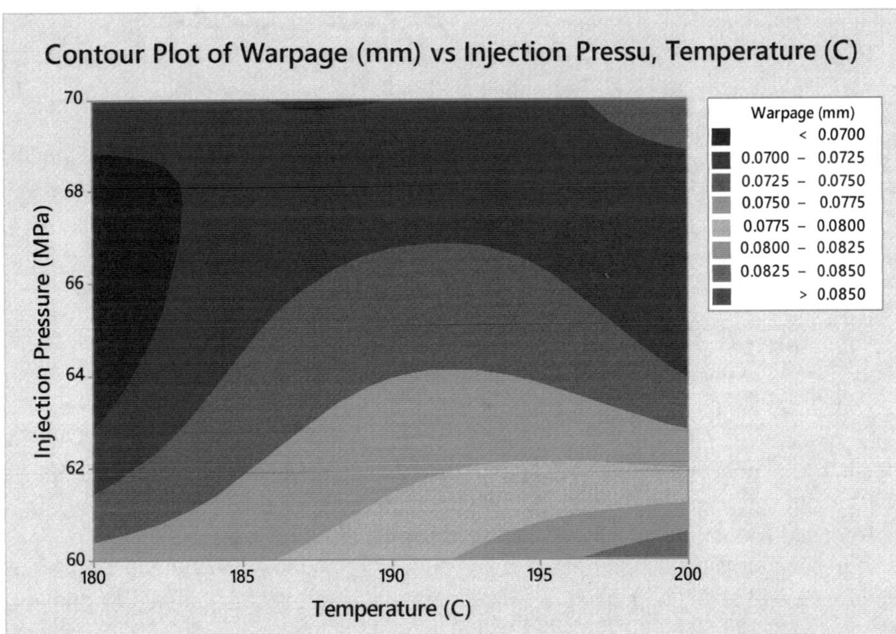

Figure 6. Graph for Warpage vs Temperature and Injection pressure for HDPE composite

The contour plot for injection pressure and temperature vs warpage is shown in fig.6. It can be observed that at 180 °C, the warpage is minimum at 63 to 69 MPa. As the temperature is increased and pressure is decreased the warpage tends to increase higher. At 200 °C the warpage is maximum and at 180 °C the warpage is minimum at all pressures comparatively.

5. Conclusion

The HDPE composite specimens were prepared by the Injection moulding process. With the parameters set of are Melting Temperature (°C), Injection pressure (MPa) and cooling time (sec). The warpage and Shrinkage values were calculated using digital vernier caliper and dial gauge. From the readings the given below conclusions are drawn.

i. As the Temperature increases the warpage value and shrinkage value also got increased, it's due to the temperature difference of the mould base and the side of plastic injection.
ii. The difference between the cold side and hot side inside the die is the main reason for causing warpage defect, and the study also shows similar effect of warpage. The injection parameters had to be controlled to avoid the development of defectsin HDPE composite.
iii. The cooling time value didn't do much influence as compared to temperature and Injection pressure parameters. The cooling time range is 15 -20 seconds at nominal pressure had better results.
iv. The Injection pressure of 63-68 MPa, the temperature range of 180- 185 °C gave better quality of specimens. The improvements in the quality will yield better results of the HDPE composite and the defects like warpage and shrinkage can be eliminated or reduced further to avoid wastage.

Further the work can be extended to find mechanical and metallurgical properties of the composites.

6. References

[1] Process–A, Injection Moulding, "*Recent methods for optimization of plastic injection moulding process–a retrospective and literature review*" (2010).

[2] Yu-Hsin Lin, Wei-Jaw Deng, Cheng-Hung Huang & Yung-Kuang Yang (2007) *Optimization of Injection Molding Process for Tensile and Wear Properties of Polypropylene Components via Taguchi and Design of Experiments Method*, Polymer-Plastics Technology and Engineering, 47:1, 96-105, DOI: 10.1080/03602550701581027

[3] Deng, Wei-Jaw, et al. "*An effective approach for process parameter optimization in injection molding of plastic housing components*" Polymer-Plastics Technology and Engineering, 47.9 (2008): 910-919.

[4] Rathi, M. M., and Mr Manoj Damodar Salunke. "*Analysis of injection moulding process parameters*" International Journal of Engineering Research and Technology 1.8 (2012): 1-5.

[5] Zhao, P., Dong, Z., Zhang, J., Zhang, Y., Cao, M., Zhu, Z. & Fu, J. (2019). *Optimization of injection-molding process parameters for weight control: converting optimization problem to classification problem*. Advances in Polymer Technology, 2020.

[6] Ramesh, M., and K. Panneerselvam. "*Mechanical investigation and optimization of parameter selection for Nylon material processed by FDM*" Materials Today: Proceedings (2020).

[7] Hernadewita, H., Rochmad, I., Hendra, H., Hermiyetti, H., & Yuliani, E. N. S. (2019). *An analysis of implementation of Taguchi method to improve production of pulp on hydrapulper milling*. International Journal of Production Management and Engineering, 7(2), 125-132.

[8] Ozcelik, Babur, and Tuncay Erzurumlu. "*Comparison of the warpage optimization in the plastic injection molding using ANOVA, neural network model and genetic algorithm*" Journal of materials processing technology 171.3 (2006): 437-445.

IOP Conf. Series: Materials Science and Engineering 1183 (2021) 012001 doi:10.1088/1757-899X/1183/1/012001

[9] Tsai, Kuo-Ming, Chung-Yu Hsieh, and Wei-Chun Lo. "*A study of the effects of process parameters for injection molding on surface quality of optical lenses*". Journal of materials processing technology 209.7 (2009): 3469-3477.

[10] Gao, H., Zhang, Y., Zhou, X., & Li, D. (2018). *Intelligent methods for the process parameter determination of plastic injection molding*. Frontiers of Mechanical Engineering, 13(1), 85-95.

[11] Sánchez-Sánchez, X., Elias-Zuñiga, A., & Hernández-Avila, M. (2018). *Processing of ultra-high molecular weight polyethylene/graphite composites by ultrasonic injection moulding: Taguchi optimization*. Ultrasonics sonochemistry, 44, 350-358.

[12] Wan Muhammad, Wan Nur Azrina, et al. "*A study of injection moulding optimum parameters to control shrinkage and warpage of polypropylene-nanoclay hinges samples*" Applied Mechanics and Materials. Vol. 465. Trans Tech Publications Ltd, 2014.

[13] Azaman, M. D., et al. "*Shrinkages and warpage in the processability of wood-filled polypropylene composite thin-walled parts formed by injection molding*". Materials & Design (1980-2015) 52 (2013): 1018-1026.

[14] Kitayama, Satoshi, Ryosuke Onuki, and Koetsu Yamazaki. "*Warpage reduction with variable pressure profile in plastic injection molding via sequential approximate optimization*". The International Journal of Advanced Manufacturing Technology 72.5-8 (2014): 827-838.

[15] Li, Dequn, et al. "*A real-time process optimization system for injection molding*" Polymer Engineering & Science 49.10 (2009): 2031-2040.

Analyzing the mechanical properties of aluminium (SiC, B_4C) hybrid Composite

V.Senthil Kannan[1], P. Pugalenthi[1], K. Lenin[2], D. Sandhiya[3], A. Sam raj Hariharan[3], S. Sakthipriyan[3], R. Ranjith Mohan[3]

[1]Assistant Professor, Department of Mechanical Engineering, Paavai Engineering College, Namakkal, 637018
[2]Professor, Department of Mechanical Engineering, K. Ramakrishna College of Engineering, Trichy, 621112
[3]UG Students, Department of Mechanical Engineering, Paavai Engineering College, Namakkal, 637018

[*]Corresponding author Email: dsandhiyadm@gmail.com

Abstract. Aluminium MMCs is play the major role in the industrial sectors. The aluminium MMCs are providing superior properties which cannot achieved by existing monolithic material. Now a days the more research is involves to the development of aluminium MMCs. Here we are choosing Al 7475 is a base metal and reinforced with SiC, B_4C by stir casting method. The different combinations of composition in volume fraction were chosen like 85%Al 7.5%SiC 7.5%B4C, 85%Al 5%SiC 10%B4C, 85%Al 10%SiC 5%B4C. The stir casted composite is machined carefully to prepare specimens of micro hardness. The mechanical properties are including the micro hardness. The stir casted composite is machined carefully to prepare specimens of micro hardness (Vickers test), tensile test, impact test (Charpy test) and micro structure as per the ASTM standards. The composition of 85%Al 7.5%SiC 7.5%B4C, was given more hardness, tensile strength and impact strength has 126.33VHN, 196.79N/mm^2 and 2 Joule. As increased the percentage of SiC is increase the mechanical properties of the Al MMC.

1. Introduction

A Composite material consist of at least two materials which will combine together to give the better properties to that of the individual constituents. There are many processes that are available to make composite materials which are extremely versatile and efficient. The end produce of the composite results in lighter, more durable and stronger solution compared to the traditional materials. In the field of design and manufacturing, the development of Composite materials is one of the most important advancement in the history of materials. The primary reason composite materials are chosen for components is because of weight for its relative stiffness and strength. Composites also exhibit greater resistance to high temperature corrosion, oxidation and wear. The composites have permeated our everyday lives such as construction medical applications, transportation, aerospace, ships and rockets, probably would not get off the ground with conventional monolithic (unreinforced) materials. The composite technology also makes the use of an entire class of solid materials is possible in application for which monolithic version are unsuited because of their great strength scalar, poor resistance to mechanical and thermal shock. Further many manufacturing processes the composite are well adapted because of its complex structures and also it reduces the manufacturing cost. Those composites are closest to our personal hygiene from our shower stalls and bath tubs made of fibre glass. Solid surface, imitation granite and cultured Marble sinks and counter tops are widely used to enhance our way of livelihood.

Content from this work may be used under the terms of the Creative Commons Attribution 3.0 licence. Any further distribution of this work must maintain attribution to the author(s) and the title of the work, journal citation and DOI.
Published under licence by IOP Publishing Ltd

IOP Conf. Series: Materials Science and Engineering 1183 (2021) 012002 doi:10.1088/1757-899X/1183/1/012002

The two constituent materials can be broken down in to two categories

1. Matrix (binder)
2. Reinforcement

Table 1. Aluminium 7475 Composition

ELEMENT	Al	Zn	Mg	Si	Cr
PERCENT %	90.3	5.7	2.3	1.50	0.22

2. Literature survey

Boopathiraman (2019) Aluminium is widely used for various mechanical application and also have excellent castability properties. But it has some limitation of pure aluminium alloys due to stiffness, wear resistance and lower strength. This limitations can be eradicated by producing hybrid metal matrix aluminium composites with the addition of reinforcement particle. This paper deals with an attempt to study the mechanical behavior of aluminium hybrid metal matrix composite. The Aluminium 7075 Metal Matrix Composite (MMC) is reinforced (5, 10, 15 Vol. %) with equal amount of Boron Carbide (B4C) and Titanium Carbide (TiC), fabricated by using stir casting method. Here the micro hardness and the ultimate tensile test is obtained at 15% like 46.10BHN & 220 Mpa[1]. T.S.A. Suryakumari (2016) In the fabrication of Al7075 metal matrix composite (MMC), stir casting method is used. In the experimental study it is revealed that the mechanical properties such as micro hardness, tensile strength and micro structure are enhanced by means of heat treatment process. The micro hardness improved by adding reinforcements to the base alloy. The addition of SiC particles improved wear properties results by the addition of AL_2O_3. Further the mechanical properties enriched by heat treatment. Micro hardness and tensile strength improved by 34% and 7% by the heat treatment process. Homogeneous mixture of the reinforcement particles with molten metal was observed in the micro structure [2]. Anil Kumar Bodukuri (2016) Powder metallurgy process is used in the preparation of aluminium, SiC, B_4C metal matrix composite. The different combination of composition in volume fraction were chosen like 90%Al 8% SiC 2%B_4C, 90%Al 5%SiC 5%B_4C, and 90%Al 3%SiC 7%B_4C. From the above combination the tap density is higher in 90%Al 5%SiC 5%B_4C, the Green density is higher in 90%Al 8% SiC 2%B_4C and the Sintered density and Hardness is higher in 90%Al 3% SiC 7%B_4C. As the percentage of boron carbide is decreasing the hardness value is also decreasing [3]. Malek Ali (2020) In this review of stir casting technique and technical challenges for ceramic reinforcement particulate and aluminium matrix composites were studies. In this process, powder form as reinforcing phases are usually distributed into mechanical stirring. Non uniform distribution and porosity in casted CMMCs are consider as the main disadvantages for stir casting process. With increasing reinforcement weight percentage, and decreasing particles size a significant improvement in the mechanical properties were observed [4]. Pradeep.P [2017] The aluminium 7075 and the reinforced material T_2B are the composites made by stir casting method. In this review the value of T_2B is 4%, 6%, 8%. Here the 8% wt of $T_2B$is cause maximum hardness 126 VHN [5]. Miss Laxmi [2017]. In aluminium 6061 SiC is used as the reinforcement material. SiC is added in 3 different composition 10%, 15%, 20%. The hardness is increased in 10% and 15%. Increase the composition of SiC increases the hardness [6]. Boopathiram C [2019] The aluminium 7075 is the reinforcement of B_4C and TiC. The B_4C and TiC is added with aluminium in 3 different composition like 5%, 10%, 15% in equal volume. The 15% of composite is results high micro hardness 46.10 VHN and ultimate tensile 220.41 Mpa [7]. Gopal Krishna U.B [2013] The aluminium with B_4C. The boron carbide is used to increase the strength. B_4C is used in 37, 44, 63, 105, 250-micron sizes separately. The 250 micron is discovered greatest for 12% wt [8].

3. Experiment

3.1. Stir Casting

The matrix material Al 7475 obtains in the form of ingots. The ingots are cleaned to remove the impurity, dust and heated in graphite crucible in induction furnace. The quantity of aluminium and the reinforcement are shown in table 2. The melting temperature was held at 780° C and the reinforcement are added in that crucible. The constant stirring of the melt carried out with alumina stirrer to get uniform distribution of the ceramic particles. The stirring was carried out for 15 minutes and the melt was poured in to the graphite mould. The stir casting process is shown in the figure 1.

The composition is chosen from the literature and the most common percentage of base metal is different. So, we are chosen the base metal is equal in all 3 compositions.

Table 2. Composition Ratio

Specimen Code	Al 7475	B_4C	SiC
1	85%	7.5%	7.5%
2	85%	10%	5%
3	85%	5%	10%

Figure 1. Stir Casting

Al 7475 **SiC** **B_4C**

Figure 2. Composite Material

Figure 3. Before Casting

Figure 4. After Casting

3.2. Wire cut EDM

Wire EDM machining is an electro thermal production process where a thin single strand metal wire, along with de ionized water allows the wire to cut through metal by the use of heat from electrical sparks, while preventing rust. virtually any conductive material can be cut using Wire EDM. This would include all metals, including steel, aluminum, brass, titanium, and alloys and super alloys of all types.

Principle of EDM: Electrical discharge machining process works on the basic principle of spark generation and metal removed by spark erosion. EDM spark erosion is same as electric spark which burn a small hole in a piece of metal through which it contacts. The spark generated by this process produces heat, which remove metal by erosion and evaporation. In this machining process both the work piece and tool must be made by conductive material. We are cutting the material for the ASTM value by using Wire cut EDM.

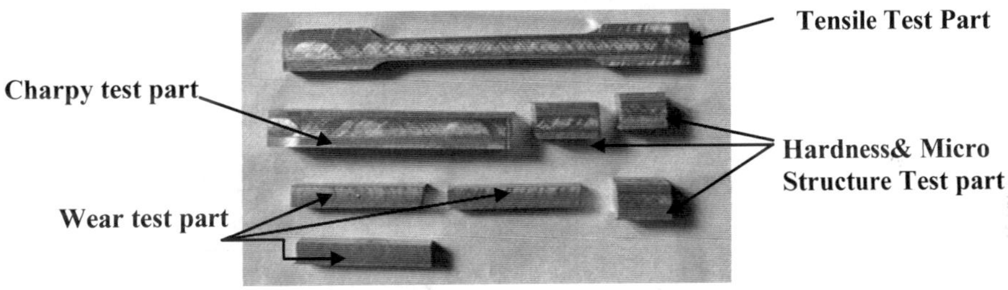

Figure 5. Material Testing

3.3. Micro structure Analysis

Micro structural analysis is used widely in industry to evaluate products and materials. Performance, response to environment and failure mechanisms are just some of the areas in which microstructural analysis can be utilized to assess and develop products.

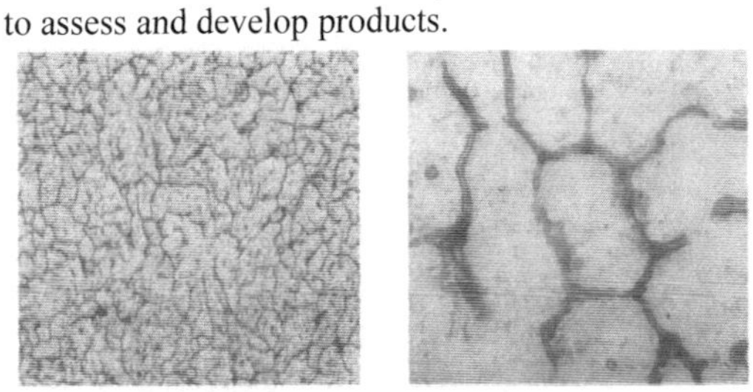

Figure 6. 85%Al 5%SiC10%B4C

IOP Conf. Series: Materials Science and Engineering 1183 (2021) 012002 doi:10.1088/1757-899X/1183/1/012002

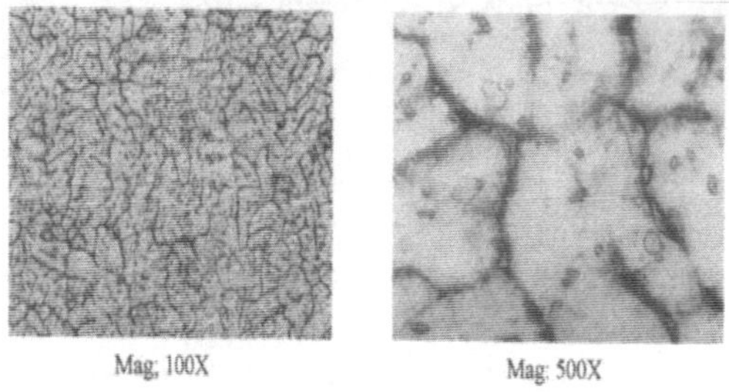

Figure 7. 85%Al 5%SiC10%B4C

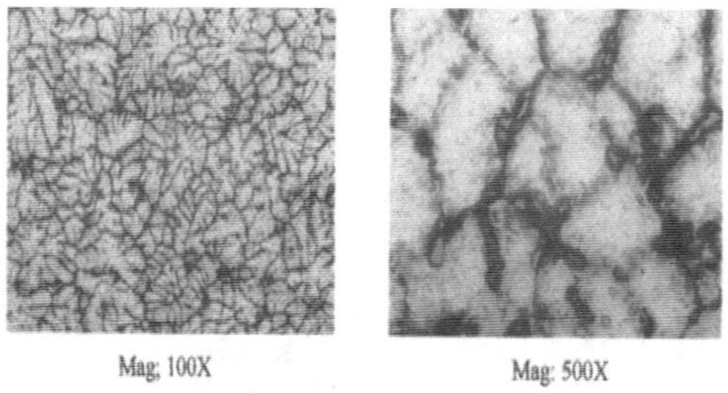

Figure 8. 85%Al 10%SiC5%B4C

4. Testing method

4.1. Vickers hardness test

Mechanical properties of stir casted material samples are determined by hardness measurement. Here we are conducting the hardness test by using micro hardness test machine.

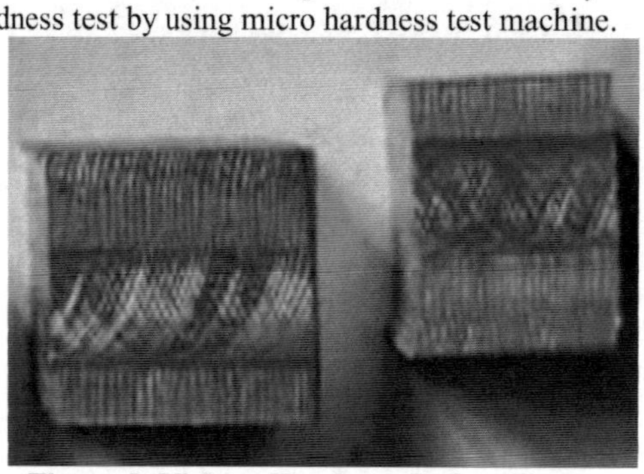

Figure 9. Vickers Hardness Before Testing

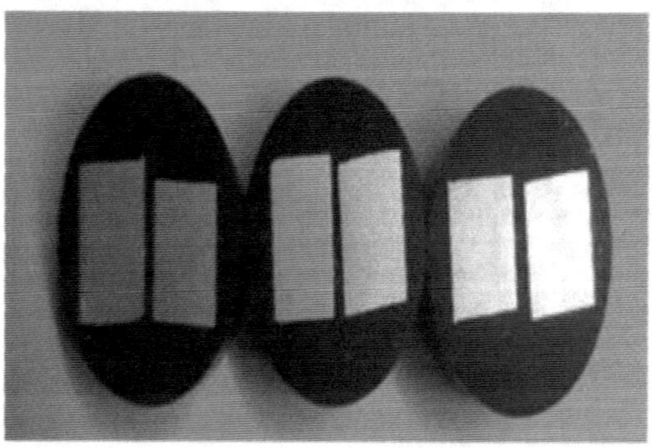

Figure 10. Vickers Hardness After Testing

Table 3. Vickers Hardness Test Result

S.NO	SAMPLE ID	OBSERVED VALUE			
		1	2	3	AVERAGE
1	01	129	124	126	126
2	02	117	121	119	119
3	03	122	119	126	122.33

4.2. Tensile test

Table 4. Tensile Test Result

Sample ID	Yield Stress(N/mm^2)	Tensile Strength(N/mm^2)	Percentage of Elongation
01	93.761	196.796	2.08
02	119.708	169.876	1.80
03	115.365	150.902	2.88

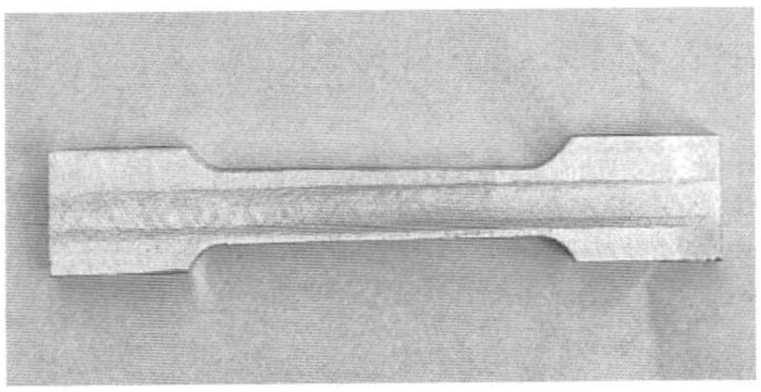

Figure 11. Tensile Test before Testing

Tensile test was performed to determine tensile properties of Aluminium Hybrid composite including yield stress and tensile strength of the specimens.

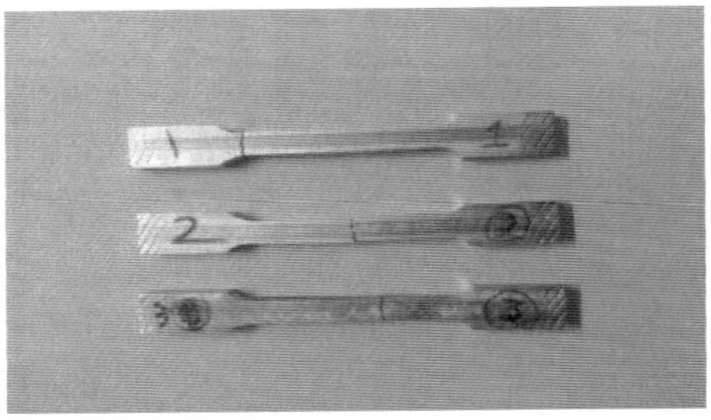

Figure 12. Tensile Test after Testing

IOP Conf. Series: Materials Science and Engineering 1183 (2021) 012002 doi:10.1088/1757-899X/1183/1/012002

Stress Vs Strain

1. 85%Al 7.5%SiC 7.5%B4C

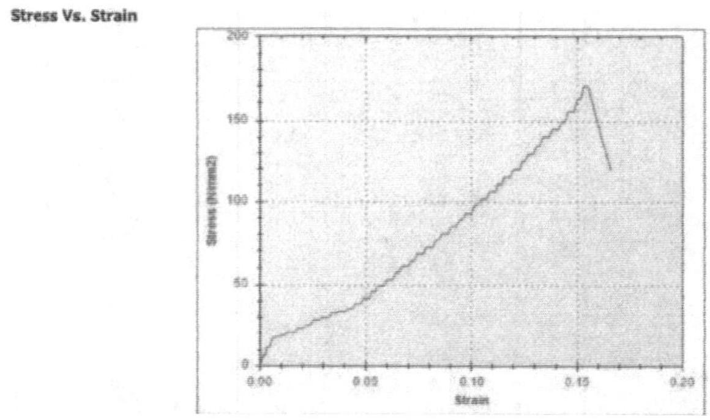

2. 85%Al 5%SiC10%B4C

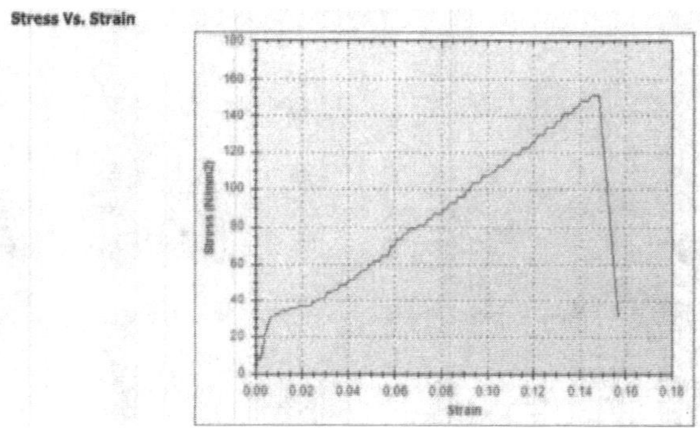

3. 85%Al 10%SiC 5%B4C

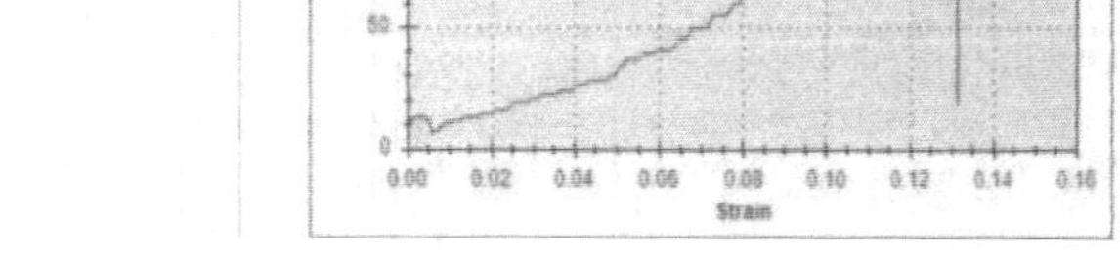

4.3. Charpy test

A Charpy test apparatus consists of a weighted pendulum, which is dropped from a specified height to make contact with the specimen. The energy transferred to thematerialcanbeinferredbycomparingthedifferenceintheheightofthependulumbeforeand after the fracture. The test specimens were prepared for 55 x 10 x 10mm. Impact values obtained by the Charpy. The Charpy test was one of the types of impact test.

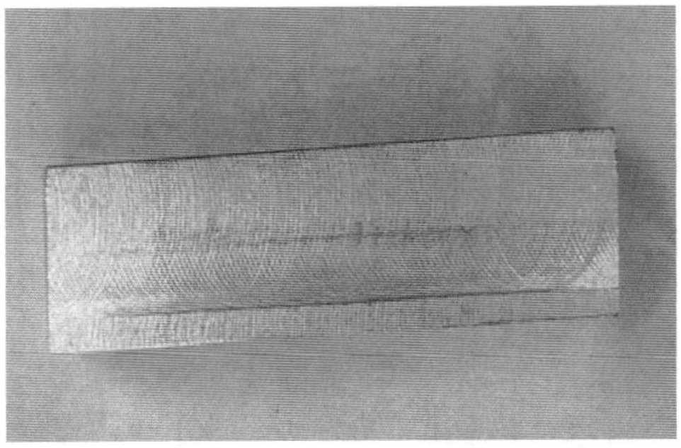

Figure 13. Charpy Material Before Test

Figure 14. Charpy Material After Test

Table 5. Charpy Test Result

Sample ID	Test Temp., °C	Charpy Impact Energy (J)
01	Room Temperature	2
02	Room Temperature	2
03	Room Temperature	2

4.4 Wear Test

Wear occurs as a natural consequence when two surfaces with a relative motion interact with each other. Wear may be defined as the progressive loss of material from contacting surfaces in relative motion. Here we are finding the wear rate in Microns.

Table 6. Wear Test Result

SAMPLE	TIME (SECONDS)		
	300	**600**	**900**
		WEAR RATE (MICRONS)	
1	259.39	478.4	770.84
2	791.67	1063.13	2187.56
3	474.11	552.03	4835.71

5. Results and discussions

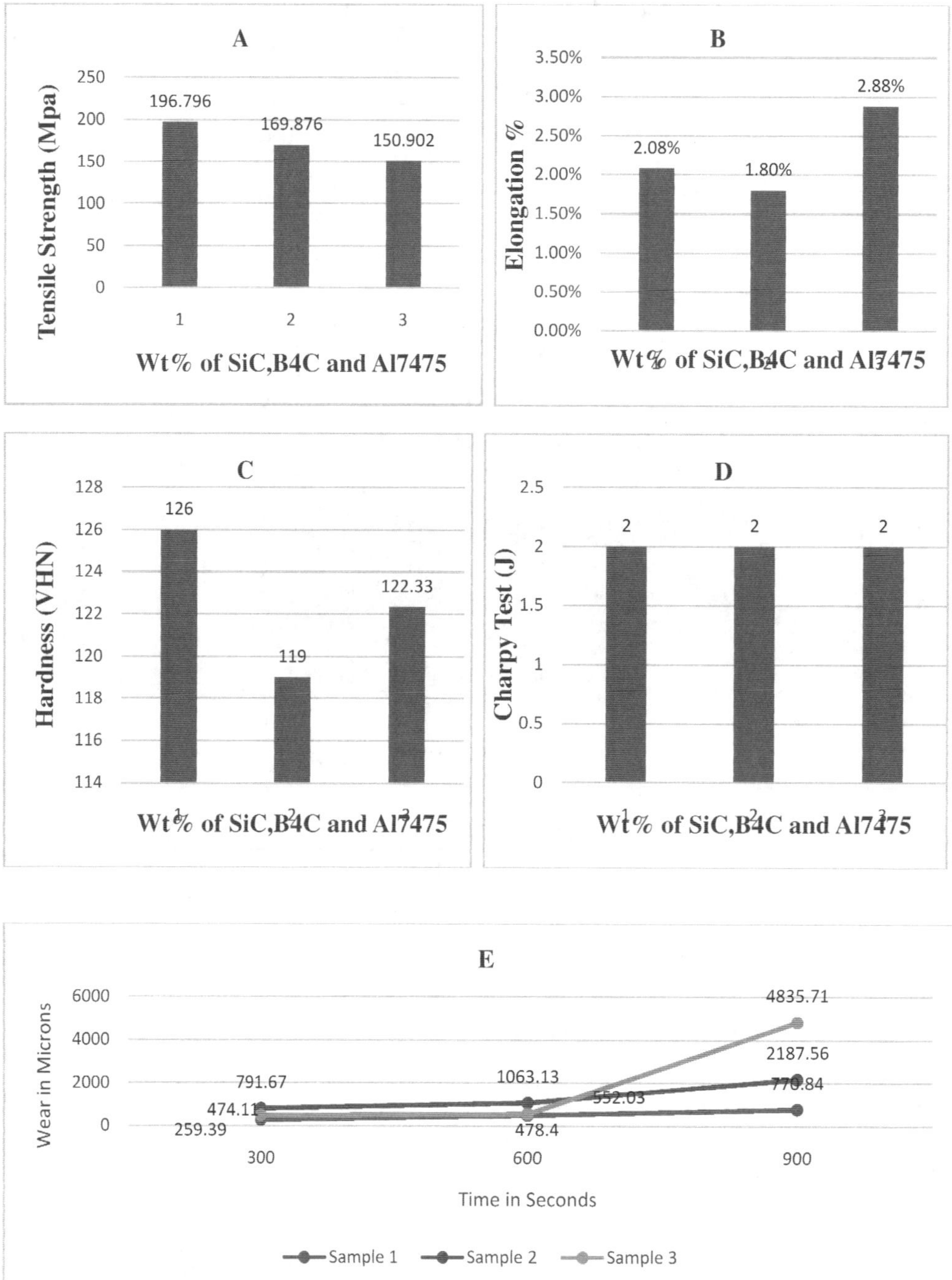

A - Tensile Strength Values of the samples
B – Elongation Values of the samples
C – Vickers Hardness values of the samples
D – Charpy Test values of the samples
E – Wear Test values of the samples

Figure 15. Results and Discussions

6. Conclusion

The important conclusion is inferred from the studies carried out on SiC, B_4C with Al7475 composite are as follows:

- The stir casting is the effective technique to produce hybrid metal matrix composites with good miscibility.
- Tensile test, Hardness test, Charpy test and Wear test have been carried out using the samples.
- With the addition of SiC increases the hardness is increases and tensile test value is decreases.
- With the addition of B_4C increases the tensile test value is increases and hardness test value is decreases
- The sample is containing 85%Al 7.5%SiC 7.5%B4Cis exhibited high tensile strength 196.796N/mm^{2,} high hardness 126 VHN and the charpy test value is 2 joules in all samples. The less wear is occurred at the 85%Al 7.5%SiC 7.5%B4C combination.

7. References

[1] Pugalenthi P, Jayaraman M, Natarajan A Evaluation of Mechanical Properties of Aluminium Alloy 7075 Reinforced with SiC and Al_2O_3 Hybrid Metal Matrix Composites 2015: 766-767

[2] Kakkar K, Jha S, Sharma R, Kamboj K. Study of the Properties of Different Aluminium Metal Matrix Composites-A Review. Proceedings of 3rd International Conference on Manufacturing, Manfex 2016:80-86.

[3] Prasad D S, Krishna A R. Production and Mechanical Properties of A356.2/RHA Composites. International Journal of Advance Science and Technology 2011;33:51-58.

[4] Vinitha, Motgi BS. Evaluation of Mechanical Properties of Al7075 Alloy, Fly ash, SiCand Red mud Reinforced Metal Matrix Composites. International Journal for Scientific Research & Development 2014;2:190-193.

[5] Pugalenthi P, Jayaraman M, Subburam V, Study of the Microstructures of and Mechanical Properties of Aluminium Hybrid Composite with SiC and Al_2O_3 2015; 1580-2949.

[6] Rahman M H, Rashed HMMA. Characterization of Silicon Carbide Reinforced Aluminium Matrix Composite. Proceedings of the 10th International Conference in Mechanical Engineering 2013:103-109.

[7] Siddique TA, Islam MT, Kabir MS, Haque MN. Effect of SiC Addition on the Indentation Hardness of as-cast Al Metal matrix composites. International Journal of Innovation and Scientific Research2014;11:433-438.

[8] Hashim J, Looney L, Hashmi MSJ. Metal matrix composites: production by the stir casting method.JournalofMaterialsProcessingTechnology1999;92:1-7.

[9] Kala H, Mer KKS, Kumar S. A Review on Mechanical and Tribological Behaviors of Stir Cast Aluminium Matrix Composites. 3rd International Conference on Materials Processing and Characterization 2014:1951-1960.

[10] Vanarotti M, Kori SA, Sridhar BR, Padasalgi SB. Synthesis and Characterization of Aluminium Alloy 356 and Silicon Carbide Metal Matrix Composite. Proceedings of 2nd International Conference on Industrial Technology and Management, Singapore 2012; 49:11-15.

[11] Saheb DA. Aluminium Silicon Carbide and Aluminium Graphite Particulate Composite. ARPN Journal of Engineering and Applied science 2011;6:41-46.

[12] Bansal S, Saini JS. Mechanical and Wear Properties of SiC/Graphite Reinforced Al359Alloy-based Metal Matrix Composite. Defense Science Journal2015; 65:330-338.

[13] Basaravaraju S, Arasukumar K, Bendigeri C, Umesh CK. Studies on Mechanical Properties and Tribological Characteristics of LM25- Graphite- Silicon Carbide and LM25-Flyash-SiliconCarbide-Hybrid MMC's. International Journal of Innovation in Science, Engineering and Technology 2012;1:107-112.

Treatment of Dye Waste Water using Moving Bed Biofilm Reactor and Granular Activated Carbon with Neem Leaf Powder

B. Sasivarman[1*], S. Dharani[2], A. Jothika[2], D. Mohana Priya[2]

[1]Assistant Professor, Department of Civil Engineering, Adhi College of Engineering and Technology, Kanchipuram, Tamilnadu, India.
[2]Department of Civil Engineering, Adhi College of Engineering and Technology, Kanchipuram, Tamilnadu, India.

[*]Corresponding author Email: sasivarman0204@gmail.com

Abstract. In recent days, the waste productions from industries are increased and direct discharge of wastes causes various pollution to the environment. Wastewater treatment in textile and dye industry mainly involves treatment of highly colored wastewater containing variety of dyes in different concentrations. The dyeing industries using different types of pigments and dyes of various concentration during their several production steps. These industries discharge a wastewater which contains organic substance, high colour with high COD and low concentration of BOD. This dye waste water are directly discharge in to the streams and thus contaminate the groundwater. To treat the wastewater, which is discharge from the dyeing industries a simple physio-chemical treatment is not sufficient for the removal of pollutants. So that a biological treatment followed with a physio-chemical treatment have been required. In this study, the experiment was carried out to reduce the organic substance, high colour, COD, BOD, and Heavy metals from dye wastewater using the laboratory scale Moving Bed Biofilm Reactor followed with Granular activated Carbon bed (GAC with NLP-MBBR) with Neem leaf powder. A various thick granular activated carbon bed was used for adsorption technique and for biological treatment of wastewater. The GAC with 8cm thick is layered followed by Ennore sand with 3 grades with 4cm thick equally. Then the MBBR was operated continuously with various retention times. The polypropylene type bio balls were used for microorganism attachment in MBBR. The outlet water from the MBBR are treated using membrane process. Membrane process using moving membranes reduces the pH, organic contents, heavy metals, COD, BOD, turbidity etc., The final effluent water is suitable for Irrigation.

Keywords: *MBBR (Membrane Bed Biofilm Reactor), GAC (Granular Activated Carbon), COD (Chemical Oxygen Demand), BOD (Biological Oxygen Demand), DO (Dissolved Oxygen), EC (Electro Chemistry), NLP (Neem Leaf Powder)*

1. Introduction

Water pollution means a deviation from pure condition, partially or completely by human activity. Major industries like pulp and paper, chemical, petrochemical, refining, metal working, food processing, textile industries etc. are the major contributors of water pollution. Textile industry generates one of the most polluting effluent compared to other industrial effluents. Due to changes in the consumer's demand, effluent contents have a rapid change. Synthetic reactive dyes are used in

Content from this work may be used under the terms of the Creative Commons Attribution 3.0 licence. Any further distribution of this work must maintain attribution to the author(s) and the title of the work, journal citation and DOI.
Published under licence by IOP Publishing Ltd

great amounts now a day. The production processes in textile industry not only utilizes large quantity of water and energy but also generate large amount of waste products. MBBR innovation is one the propelled strategy for the treatment of waste water. The microscopic organisms in the biofilms will devour the organics and supplements introduce in the waste water. It is hard to encourage the water straightforwardly to the MBBR tank. So a pre-treatment of waste water is required. So what we have picked granular actuated carbon. This GAC will diminish the COD level; hence the microscopic organisms connected in the biofilms won't be bothered. The GAC have high retaining power. It expels the shade of the waste water at first. The primary favorable position of the MBBR tank is they diminish the aggregate broke up solids viably. As the microscopic organisms expend its nourishment from the waste the broke up solids can be evacuated effectively. Aside from that, it additionally diminishes pH, COD, BOD, TSS, Alkalinity, Hardness, turbidity, and so forth. The powder of these leaves can replace the chlorine used as a sterilizer. The results of this study indicated that the powder and the leaves of the Neem tree may be used as an alternative to chemicals that have environmental side effects and that the wastewater can be re-used safely.

2. Sample collection and preparation

In our adventure we have used the GAC- Neem and Sand filter tank as primary process and MBBR tank as secondary process. The raw dye waste water was collected at the quantity of 200litres. In GAC Filtration tank, the granular activated carbon made as a layer of bed with the stratum of up to 25cms. The GAC also remove the organics substances, heavy metals, etc. The second process is MBBR tank. The MBBR tank comprises of aerators and biofilms. The biofilms are the polypropylene materials where the bacterium can be growth effectively. Initially the biofilms are soaked in dairy animals excrement in a container. It is then blended with the activated sludge gathered from the air circulation tank of STP. This seed slime will start the development of microorganisms in the biofilms. Initially the waste water is poured in the GAC filtration tank and permitted to dormant for 12 hours. The outlet of GAC filtration tank is connected with MBBR tank. The capacity of tank is 25 litres the quantity of biofilms taken as 100. By using the aerators the tank is fully aerated. This helps to lift the biofilms and make them to turn without making them static. The best possible maintenance time is noted. The maintenance time are taken as 2hours, 4hours,6 hours, 8 hours.

3. Experimental study

3.1. Initial characteristics of waste water

The textile effluent is to be collected and that effluent initial characteristics are completed. The chemical characteristics were completed. There is pH, Turbidity, Total Solids, Total Dissolved Solids, Ammonia, Conductivity, Calcium, Hardness, Total Suspended Solids, BOD, COD, and DO. The obtained results are shown in the Table1.

Table 1. Initial characteristics of tannery waste water and activated sludge.

S. No	Initial characteristics	W. W	Unit
1	Ph	8.76	-
2	Ec	2.65	Mhos/cm
3	Cl	349	mg/l
4	Cod	1830.8	mg/l
5	Bod	359	mg/l
6	Do	50	mg/l
7	Alkalinity	1000	mg/l

IOP Conf. Series: Materials Science and Engineering 1183 (2021) 012003 doi:10.1088/1757-899X/1183/1/012003

8	Hardness	168.75	mg/l
9	Ca Hardness	70	mg/l
10	Mg hardness	98.7	mg/l
11	Sulphate	1687.5	mg/l
12	Acidity	925	mg/l
13	Turbidity	100	mg/l
14	Ts	3000	gms

3.2. GAC
Granular activated carbon (GAC) is a hybrid mixture of a wide variety of graphite platelets that are interconnected by non-graphitic carbon bonding. The adsorptive capacity of GAC makes it ideal for removing a variety of contaminants from water, air, liquids and gases. GAC is also an environmentally responsible product that can be reactivated through thermal oxidation and used multiple times for the same application. Figure 1 shows Granular Activated Carbon.

Figure 1. Granular Activated Carbon

4. GAC-Neem filter process
The capacity of tank is up to 6 liters and 25cm height. The various layers of GAC-NEEM and sand are prepared. The 3 grades of ENNUR sands are layered continuously each up to 3cm. Then the NEEM powder and GAC is layered up to 8cm. The surface loading rate of activated carbon is low. Lower SLR and longer in adsorbent limit will increase adsorption process. The higher volume of the contact bed increase yield. In connection to the qualities of GAC, it might create a superior expulsion of color, odour, heavy metals etc. The expulsion of adsorption treatment was discovered utilizing 8 cm of GAC with 90% evacuation of COD, 60% evacuation of smelling salts and 58% expulsion of shading. The GAC treated water is directly discharged in to the MBBR tank. The final effluent is to be meet the water system norms. Figure 2. shows the GAC-Neem filter tank.

Figure 2. GAC-Neem filter tank

IOP Conf. Series: Materials Science and Engineering 1183 (2021) 012003 doi:10.1088/1757-899X/1183/1/012003

5. MBBR tank

The MBBR system consists of an aeration tank (similar to an activated sludge tank) with special plastic carriers that provide a surface where a biofilm can grow. The carriers are made of a material with a density close to the density of water (1 g/cm^3). An example is high-density polyethylene (HDPE) which has a density close to 1 g/cm^3. The carriers will be mixed in the tank by the aeration system and thus will have good contact between the substrate in the influent wastewater and the biomass on the carriers. To prevent the plastic carriers from escaping the aeration it is necessary to have a sieve on the outlet of the tank. To getting away from air circulated biofilms it is important to have a strainer on the outlet of the tank. Figure 3. shows MBBR tank.

Figure 3. Membrane Bed Biofilm Reactor tank

6. Treatment in MBBR tank

In this biological treatment, Poly Vinyl Chloride (PVC) transporter materials were used as a part of MBBR. Carrier media's of MBBR are nutrient with (Micro bacterium Marnilacus), diary animals discrete.

6.1. Biocarriers

A biofilm used to gather of microorganisms in which cells adhere to each other and frequently additionally to a surface. These follower cells end up inserted inside a foul extracellular network that is made out of extracellular polymeric substances (EPS). Total of microorganisms in which cells that are much of the time installed inside a self-created grid of extracellular polymeric substances (EPSs) stick to each other and additionally to a surface. Figure 4. shows MBBR Treatment Setup

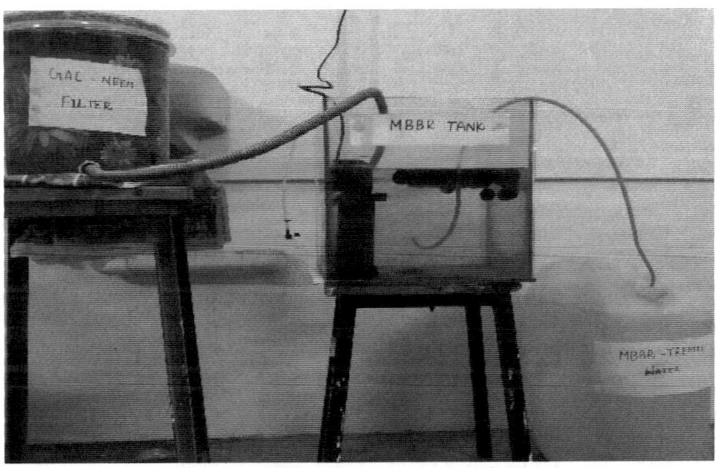

Figure 4. MBBR Treatment Setup

7. Reuse of Dye waste water in Agriculture

Using treated dye waste water effluents and sludge on agricultural land provides an alternative to disposal by utilizing the recyclable constituents in sludge and waste water in the production of crops. Figure 5. shows Planting Spinach Using Dye Waste Water.

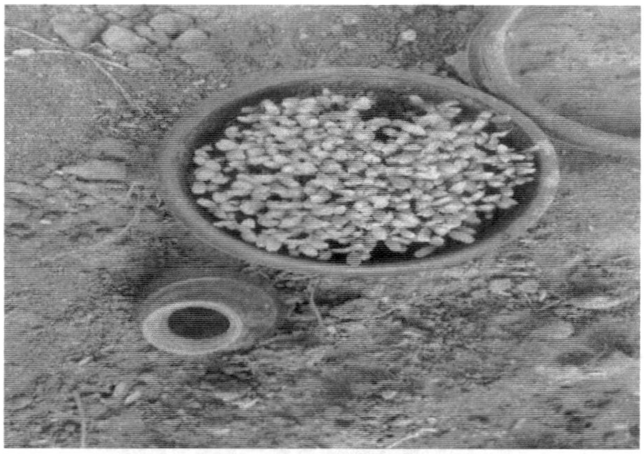

Figure 5. Planting Using Dye Water

Figure 6. shows planting of spinach using GAC treated water.

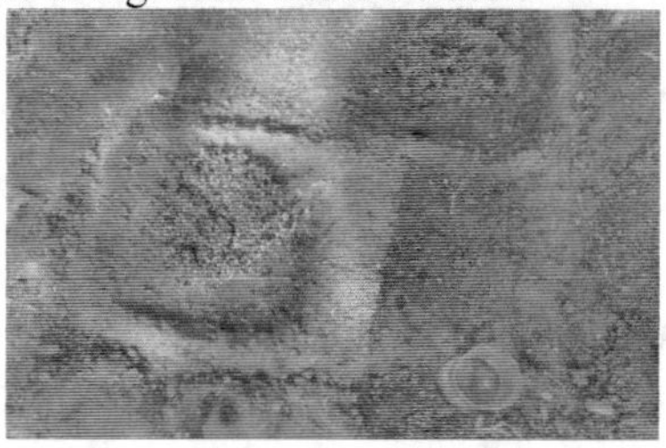

Figure 6. planting using GAC treated water

Figure 7. shows planting of spinach using MBBR treated water.

Figure 7. planting using MBBR treated water

8. Results and discussion

The wastes associated are decreased and evacuated according to the natural standards. The treated water can be utilized for the water system. The table 2 shows Result of GAC & 3 shows Result of MBBR treated value figure 8,9,10 &11 shows the results of pH, Chloride, COD& BOD. The BOD, COD of water also decreased 88%.A biofilm was used to gather of microorganisms and surface loading rate of activated carbon is low. Lower SLR and longer in adsorbent limit will increase adsorption process. The expulsion of adsorption treatment was discovered utilizing of GAC with 90% evacuation of COD. The NEEEM and GAC play the vital role for reducing a BOD and COD. In most of the paper almost they had used whether carbon or neem powder but in our project we introduced both GAC and NEEM powder to get a better result. The total solids evacuated are almost 80%. The treated water meets future water demand. The treated water is to be safe for irrigational purpose.

Table 2. Results of GAC treatment

S.No	Parameter	GAC Outlet	Unit
1.	pH	7.75	-
2.	EC	2.32	Mhos/cm
3.	Cl	134.9 mg/l	mg/l
4.	COD	570.4 mg/l	mg/l
5.	BOD	30.30 mg/l	mg/l
6.	DO	25 mg/l	mg/l
7.	Alkalinity	750 mg/l	mg/l
8.	Hardness	54 mg/l	mg/l
9.	Ca Hardness	30 mg/l	mg/l
10.	MgHardness	24 mg/l	mg/l
11.	Sulphate	926.1 mg/l	mg/l
12.	Acidity	420 mg/l	mg/l
13.	Turbidity	35 mg/l	mg/l
14.	TS	800 mg/l	Gms

Table 3 shows the Results of MBBR treatment

Table 3. Results of MBBR treatment

S.No	Parameter	MBBR outlet			
		2hr	4hr	6hr	8hr
1.	pH	6.2	6.7	7.2	7.5
2.	EC	1.5	1.52	1.7	1.73
3.	Cl	89.9	90.7	92.9	96.3
4.	COD	138	128.8	165.6	174.8
5.	BOD	9.09	6.06	3.03	3.51
6.	DO	15	13.5	12	12
7.	Alkalinity	475	525	550	575
8.	Hardness	55	57.5	60	62.5
9.	Ca Hardness	18	18	20	20
10.	Mg Hardness	37	39.5	40	42.5
11.	Sulphate	643	660	687	690

12.	Acidity	125	122	124	132
13.	Turbidity	48	51	56	58
14.	TS	318	331.5	342	345

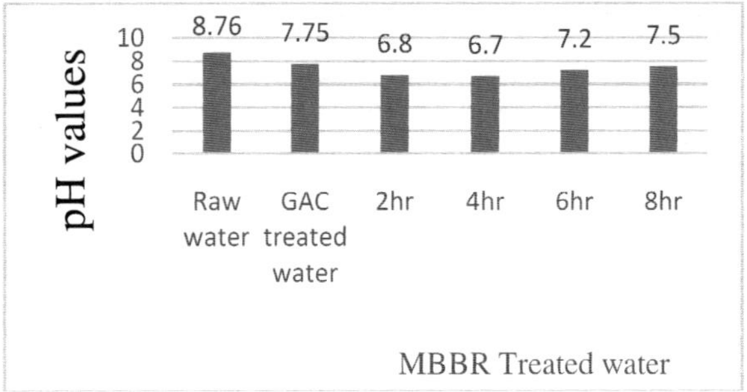

Figure 8. Values of pH

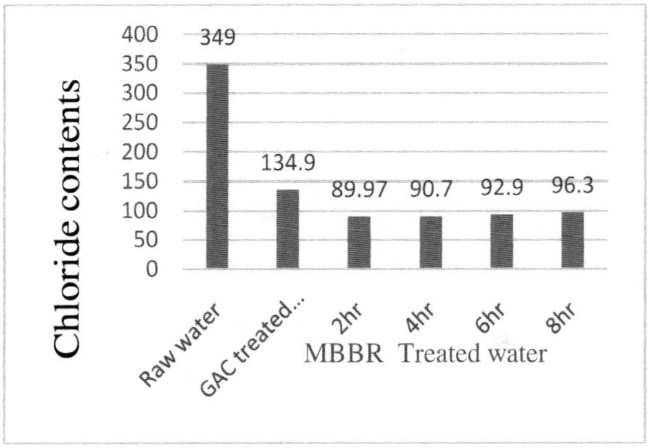

Figure 9. Values of Chlorides

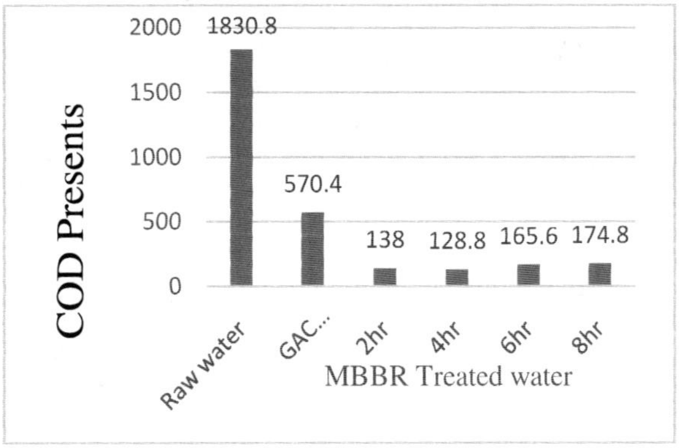

Figure 10. Values of COD

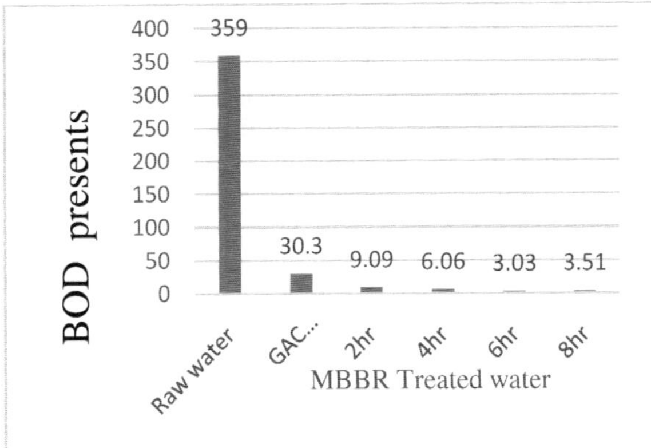

Figure 11. Values of BOD

9. Conclusion

- This experiment has demonstrated that the Moving Bed Biofilm process alongside Granular Activated Carbon can be utilized as a perfect and effective choice for the organics and supplement expulsion from color squander water.
- The physical and compound attributes of the wastewater from the coloring business are especially decreased and it can be utilized for water system reason.
- The estimation of pH is almost 7.5 at ideal maintenance time.
- The COD evacuated about 88%. The BOD and DO is to be evacuated about 30.3& 50% at primary process itself. The shade of the wastewater is totally evacuated.
- The solids which lessened almost 80% at the primary procedure itself.
- The chlorides removal is done almost 93%.
- The alkalinity of water removed 82%.
- The hardness present in the water is removed up to 42%.
- The turbidity of water is evacuated almost 60%.
- Our MBBR-GAC can evacuate the physical and compound attributes of the wastewater from the coloring effectively. In this manner it can be utilized for water system.

10. References

1. Ramesh Babu, R., Parande A.K. and Raghu, S. Textile innovation: Cotton Textile Processing: Wastewater Generation and Effluent Treatment. Diary of Cotton Science 2 (2001) 141-153.
2. John Brinkley, Chantler H. Johnson, Robert Souza, Moving Bed Biofilm Reactor Technology–A Full-scale establishment for treatment of Pharmaceutical wastewater 223 (2002) 152–161.
3. K.Hunger, et.al, Industrial dyes: chemistry, properties, applications, Wiley-VCH, Weinheim, Cambridge, 151 (2003) 316–322.
4. Surreyameric et.al, Reactive dyes removal from wastewater by combined advanced treatment Journal of industrial and engineering chemistry. 147 (2005) 297–306.
5. Churtie. R.M, et.al, Environmental aspects of textile dyeing, wood head, bocaraton, Cambridge 136 (2007) 800–808.
6. Ramesh babu.R, et.al, Adsorption and kinetic studies of methylene blue on zeolite synthesized from fly ash, Desal. Wat. Treat. 99 (2007) 5368–5373.
7. Gupta.V.K, et.al, Application of low cost absorbents for dye removal, Environmental management. 69 (2009) 196–203.

8. Sivamani.S, et.al, Wastewater treatment with activated carbon, Achieves of environmental protection. 78 (2009) 307–311.
9. C.I.Pearce, J.R.Lloyd, J.T.Guthrie and et.al, The removal of color from textile wastewater using whole bacteria cells. 135 (2010) 174–184.
10. Bae, W., Han, D., Cui, F. and Kim, M. Microbial evaluation for biodegradability of recalcitrant organic in textile wastewater using an immobilized-cell activated sludge process. KSCE Journal of Civil Engineering. 201 (2010) 191–199.
11. Wessman, F.G. and Siljudalen, J.G. Rusten, B., Yuegen, E.Y., Qiaogeng, Z., Guanghui, H., Expanding the limit with respect to treatment of substance plant wastewater by supplanting existing suspended bearer media with Kaldnes Moving BedTM Media. Control Board (Ministry of Environment and Forests, Govt. of India), Water Environment Federation, 116 (2010) 211–217.
12. T.Robinson, G.Mcmullan, R.Marchant, P.Nigam, Remedition of dyes in textile effluent: a critical review on current treatment technologies with a proposed alternatives, Bioresour, Technology. 41 (2010) 487–495.
13. Dash.B, et.al, Competitive adsorption of dyes on activated carbon, B.Tech thesis 113 (2010) 81–88.
14. S. Chen, J. Zhang, C. Zhang, Q. Yue, Y. Li, C. Li,et.al, Equilibrium and kinetics studies of methyl orange and methyl violet adsorption on activated carbon derived from phragmitesaustralis, Desalination 27 (2011) 1129–113
15. G.Crini, Non-conventional low-cost adsorbents for dye removal: a review, Bioresour, Technol.97(2011) 1061-1085.
16. Mittal.A, Malviya.A, Kaur.D, Mittal.J, Kurup.L, Studies on the adsorption kinetics and isotherms for the removal and recovery of Methyl Orange from wastewaters using waste materials 69 (2011) 1151–1158
17. Desai, P.A. and Kore, V.S. Performance evaluation of effluent treatment plant for textile industry in Kolhapur of Maharashtra. Universal Journal of Environmental Research and Technology 1 (4) (2011) 560-565.
18. Schneider.E, et.al., MBBR assesment for oil refinery wastewater treatment with post ozonation and BAC for wastewater reuse, Water science and technology 39 (2011) 1347–1353
19. Syafalni, S., Abustan, I., Dahlan, I., Wah, C.K. and Umar, G. Treatment of dye wastewater using granular activated carbon and zeolite filter. Modern Applied Science 6 (2) (2012).
20. Vikasdinkar, crosavi.F, et.al, Textile wastewater treatment by chemical coagulation technology, Textile wastewater treatment by chemical coagulation technology 158 (2013) 142–150.
21. Abdul, A., Farhana, T. and Mohd, A. Essential Design of a Fluidized Bed Reactor for Waste water Treatment utilizing Fenton Oxidation. International. Diary of Innovation Management and Technology 5 (2) (2014) 187-200.
22. Mona.A, Sherief.F, et.al., Treatment and characterization of wastewater from various dyeing industries using different adsorbents,Journal of water resource engineering 171 (2015) 54–60
23. Gayathri.K, Viji.R, et.al., Use the neem leaves for colour removal from wastewater, Journal of Engineering and Applied Science 175 (2016) 1–11.
24. Deepika Patel and et.al Adsorption of malachite green on groundnut shell waste based powdered activated carbon, Waste Manage. 77 (2017) 247–255.
25. Mohamed.A, Ahmed.E, and et.al., Cationic and anionic dye adsorption by agricultural solid waste, Journal of scientific research 98 (2017) 2369–2385.
26. Vaidhegi.S, Ajithselvam.A, and et.al., Treatment of dye wastewater using MBBR reactor, Journal of Advanced in Dynamical and Control Systems 9 (2018) 105–111.
27. Anupama ranjini. K, et.al., Characterisation of grey water and treatment using MBBR reactor, Environmental Management Journal 11 (2019) 288–293.
28. IS 10500:1999, water Standards for drinking and various purposes.

IOP Conf. Series: Materials Science and Engineering 1183 (2021) 012004 doi:10.1088/1757-899X/1183/1/012004

Metal-insulator transition in Cr-doped hollandite vanadate $K_2V_8O_{16}$

S Biswas[1, 2] and M De Raychaudhury[2*]

[1]Department of Physics, Barasat Government College, Kolkata-700124, India
[2]Deptartment of Physics, West Bengal State University, Kolkata-700126, India

*Email: molly.deraychaudhury@gmail.com

Abstract. The hollandite $K_2V_8O_{16}$ is a metal with ferromagnetic spin fluctuations in the high temperature (tetragonal I4/m) phase and encounters metal-insulator transition (MIT) at 170 K. The mechanism behind the MIT is still controversial due to inadequate information on the electronic structure. Furthermore, substitution of Rb for K or Ti for V increases the transition temperature. We have investigated whether break in the mirror symmetry is responsible for the insulating ground state of $K_2V_8O_{16}$ and since the Cr counterpart is a robust ferromagnet, whether Cr doping which leads to break in mirror symmetry can help in driving a gap. We find that both 25% and 75% Cr doping break the mirror symmetry but only the former leads to an insulating ground state. Ferromagnetism (FM) remains intact in the insulating phase of $K_2V_6Cr_2O_{16}$. The structural, electronic and magnetic properties of pure and doped $K_2V_8O_{16}$ were investigated within first-principles calculations using density functional theory (DFT). Electron correlation suppresses orbital fluctuations between the partially occupied Cr and V-3dt_{2g} states. Consequently, transfer of charge (electron) from V-3d to Cr-3dt_{2g} states is observed which facilitates Cr^{3+}-V^{4+} charge ordering. Furthermore, Peierls like structural distortion is associated to the breaking of mirror symmetry in the Cr-V rectangular four chain columns within the crystal. Therefore, the simultaneous effect of Peierls instability, charge ordering and Coulomb correlation is responsible for the MIT in $K_2V_6Cr_2O_{16}$. Besides, Cr-3d O-2p hybridization and Cr-O coupling increases with Cr doping. These two effects are cooperatively responsible for the observed FM in insulating $K_2V_6Cr_2O_{16}$. More interestingly, the strength of FM is augmented with Cr concentration in $K_2V_8O_{16}$.

1. Introduction

In recent years, studies on electronic and magnetic properties of hollandite crystals have attracted considerable interest in the field of strong electron correlations due to their exotic phenomena such as metal-insulator transition (MIT) as a function of temperature or the degree of doping level. The latter is generally controlled by the degree of chemical substitution. The hollandite oxides [1] have general formula $A_xB_8O_{16}$ (with $0 \leq x \leq 2$), where A represents large tunnel ions such as Na, K, Ba, Sr, Rb, Cs and B represents cations such as Mn, Ti, Al, Cr, V, Si. The hollandite $K_2Cr_8O_{16}$ is a paramagnetic metal (PM) at room temperature, encounters first step transition from PM to ferromagnetic metal (FM) at 180 K and finally to a ferromagnetic insulator (FI) below 95 K with a supercell structure (monoclinic I2/m) [2, 3]. The ferromagnetism (FM) in the insulating phase is very surprising because it is usually accompanied by metallicity and insulating transition metal oxides are normally anti ferromagnetic [4]. Several explanations have been offered to explain the mechanism of MIT in $K_2Cr_8O_{16}$ [5-8]. Some groups demonstrated that MIT is due to a unique charge ordering [5] or structural distortion due to the formation of Cr- tetramers running in the c-direction [6-8]. In Ti-series,

Content from this work may be used under the terms of the Creative Commons Attribution 3.0 licence. Any further distribution of this work must maintain attribution to the author(s) and the title of the work, journal citation and DOI.
Published under licence by IOP Publishing Ltd

$K_xTi_8O_{16}$ has been reported to be either a dilute ferromagnetic semi-conductor [9] or a paramagnetic metal, depending on the carrier concentration. In the Mn counterpart $K_xMn_8O_{16}$, a variety of structural, electronic and magnetic phase transitions have been reported [10, 11]. The concentration of K-ions in this compound is reduced and the actual composition tends toward $K_{1.6}Mn_8O_{16}$ [12]. The crystal structure of $K_2V_8O_{16}$ also belongs to the family of hollandite oxides, where one dimensional (1D) double strings of edge shared VO_6 octahedron forms a V_8O_{16} framework that is assumed to contain V ion in a mixed-valent state +3.75. The crystal contains two planes of V atoms in a unit cell; one is at the topmost plane and the other is at the bottom plane. Any V atom of the topmost plane has two neighbouring V atoms at the bottom plane. One of these two V atoms is linked with the topmost V atom by edge-shared oxygen (O1), while the other is connected by corner-shared oxygen O2) as can be seen in figure 1(a).

The hollandite vanadate $K_2V_8O_{16}$ is a metal with ferromagnetic spin fluctuations [13, 14] at the high temperature (HT) tetragonal I4/m structure. It is a mixed valent oxide, which attracts strong attention due to its MIT. The average valence of V atoms in this compound is +3.75 ($V^{3.75+}$). Therefore, the system is in the charge-ordered state of V^{3+}($3d^2$): V^{4+}($3d^1$) 1:3. Isobe et al. [14] synthesized a hollandite vanadate $K_2V_8O_{16}$ by a solid-state reaction method under high pressure and investigated its structural, electronic and magnetic properties. They found that $K_2V_8O_{16}$ is a ferromagnetic metal at room temperature with tetragonal I4/m symmetry. It encounters MIT at T_{MI} = 170 K with a structural phase transition to monoclinic $\sqrt{2}a\times\sqrt{2}b\times2c$ [16]. Different models have been discussed to explain the characteristics of $K_2V_8O_{16}$ in the insulating phase [15-20]. A recent study reported that half of the V chains are dimerized perfectly, which further form spin singlets that is responsible for the suppression of magnetic susceptibility [16]. Nevertheless, mechanism behind the MIT is still controversial due to inadequate information on the electronic structure (densities of states, band structures). Furthermore, substitution of Rb for K or Ti for V increases the transition temperature (T_{MI}). For example, Rb substitution continuously increases T_{MI} and it reaches 220 K for $Rb_2V_8O_{16}$ [8]. Furthermore, a slight amount of Ti-substitution for V atoms raises the transition temperature up to 270 K at x=0.5 in $K_2V_{8-x}Ti_xO_{16}$ from 170 K at x= 0 [9]. However, there have been no reports on the electronic structure till date to realize the electronic structure of Rb or Ti substituted $K_2V_8O_{16}$. Besides, reports on Cr substitution effect on the structural, electronic or magnetic properties of the present material are not available.

$K_2V_8O_{16}$ is not a robust ferromagnet, whereas $K_2Cr_8O_{16}$ is so. Furthermore, $K_xTi_8O_{16}$ (x=1.43) is a FM [21]. However, T_{MI} goes on increasing as Cr is replaced by V and then V is replaced by Ti. Our target is to keep ferromagnetism (FM) intact and simultaneously keep T_{MI} near room temperature. This is challenging. In order to test whether FM is robust or not, one has to compare the total energies of particular compound in paramagnetic (PM), FM and antiferromagnetic (AFM) configurations. In case of the compounds under consideration, one has to compare PM and FM energies. Now T_{MI} has not been calculated and it is not possible for us to do that. So our motivation was to assess whether break in the mirror symmetry is responsible for the insulating ground state of $K_2V_8O_{16}$ and since the Cr counterpart is a robust ferromagnet, whether Cr doping which leads to break in mirror symmetry can help in driving a gap. We find that that both 25% and 75% Cr doping break the mirror symmetry but only the former leads to an insulating ground state. So it would be interesting to see whether robust FM will be observed in this insulating ground state or not and we have explored that FM is augmented markedly with Cr doping.

2. Method of Calculations

In this paper, we performed first-principles electronic structure calculations based on the density functional theory (DFT) [22] as implemented on the linearized-muffin-tin orbital (LMTO) method [23]. The majority spin electrons are metallic whereas minority spin electrons are semiconducting with a band gap. The local density approximation (LDA) is used for the exchange and LDA+U [24], with the inclusion of Hubbard type Coulomb interaction U=4 eV for electron correlations. It should be noted that Kim *et al.* [19] adopted $U_{eff}=U-J$ =3 eV on the V-3d electrons to describe the insulating ground state of $K_2V_8O_{16}$ in its low-temperature (LT) phase, whereas Toriyama *et al.* [17] used U=6 eV. Considering the above cases, we have applied an intermediate U value of 4 eV on the V-3d electrons of tetragonal $K_2V_8O_{16}$. The spin polarized calculations with and without U have been performed. The unit cell is tetragonal body centred in the space group symmetry I4/m. The primitive unit cell contains one K-sites, four V-sites and eight O-sites i.e. KV_4O_8. The lattice constants are a=b=9.963 Å and c=2.916 Å [25]. Then, spin-polarized self-consistent calculations were carried out for 25% (the corresponding compound $K_2V_6Cr_2O_{16}$), 50% (the corresponding compound $K_2V_4Cr_4O_{16}$) and 75% (the corresponding compound $K_2V_2Cr_6O_{16}$) Cr-doping, all including electronic correlations.

3. Results and discussion

3.1. Structural investigations

We first investigated the high-temperature structures of pure and doped $K_2V_8O_{16}$ for three different doping levels of Cr at V sites. For $K_2V_8O_{16}$, we observed four V and four corner-shared oxygen (O2) atoms form rectangular four chain columns (RFCCs), running in the c-direction [see figures 1(a) and 2(a)]. Two consecutive RFCCs in the ab plane are linked by edge-shared oxygen atoms (O1). In each RFCC, two V atoms are located at one opposite diagonal positions (z = 0 or 1), whereas the other two V atoms are located at the other opposite diagonal positions (z=1/2).

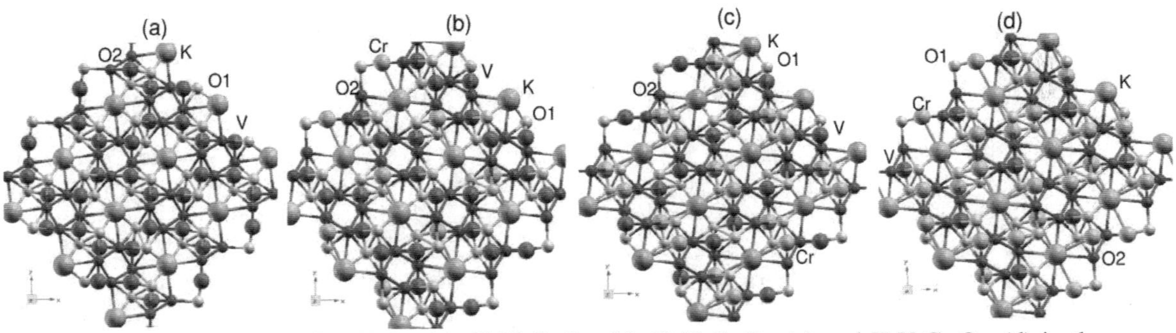

Figure1. Crystal structure of $K_2V_8O_{16}$ (a), $K_2V_6Cr_2O_{16}$ (b), $K_2V_4Cr_4O_{16}$ (c) and $K_2V_2Cr_6O_{16}$ (d) in the high-temperature tetragonal I4/m phase. The green, violet, cyan, yellow, red solid spheres represent K, V, Cr, O1 and O2 atoms respectively.

As a result, every mirror symmetry is protected in the ab plane, which is responsible for uniform V-O bond distances which further results in uniform V-V distances in the double chains running in the c-direction. Next, we concentrated on the structure of $K_2V_6Cr_2O_{16}$. The primitive unit cell of $K_2V_6Cr_2O_{16}$ consists of one K and Cr atom each, three V, four each of O1 and O2 atoms. The single Cr, three V and four O2 atoms form an RFCC, found in ab plane and running in the c direction. In each RFCC, two V atoms are found at one opposite diagonal positions (z = 0 or z = 1), whereas the Cr and another V atoms are found at the other opposite diagonal positions (z = 1/2). However, every mirror symmetry is preserved by the V atoms at z=0 or z=1 plane. Thus, V-O2 distances in the c-direction are uniform, which further results in uniform V-V distances in the double chains. Besides, Cr and V atoms at z =1/2 plane are shifted from the mirror plane. Indeed this symmetry breaking will result in long/short Cr-Cr and V-V distances in the double chains running in the c-direction as can be seen in low-temperature monoclinic I2/m structure of $K_2V_8O_{16}$ [16-19]. These bond alternations result in the formations of Cr-Cr or V-V dimers in the double chains. The formation of dimers in the Cr/V double chains further results in Peierls instability within the crystal.

The primitive unit cell of $K_2V_4Cr_4O_{16}$ consists of one K, two each of V and Cr, four each of O1 and O2 atoms. In this structure, Cr, V and four O2 atoms form an RFCC in the ab plane [see figures 1(c) and 2(c)]. In each RFCC, two Cr atoms at one opposite diagonal positions and two O atoms are located at the height z = 0 or z = 1. Also, two V atoms at the other opposite diagonal positions and the other two O2 atoms are located at the height z=1/2 plane [see figure 2(c)]. Therefore, every mirror symmetry is protected in each RFCC. Consequently, all of the Cr-Cr and V-V distances in the double chains remain uniform, resulting in the non-formation of dimers in the Cr/V double chains.

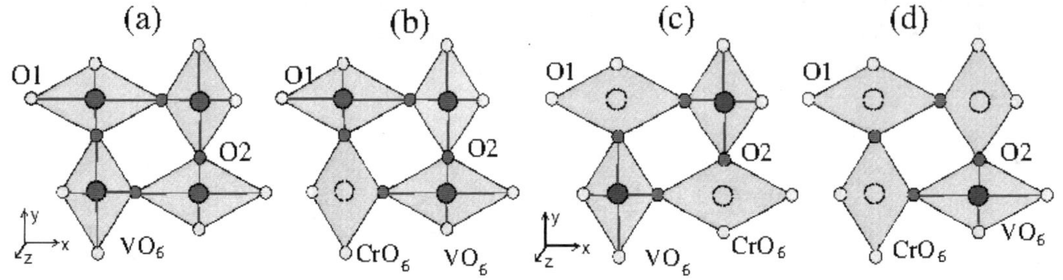

Figure 2. Schematic illustrations of RFCCs in $K_2V_8O_{16}$ (a), $K_2V_6Cr_2O_{16}$ (b), $K_2V_4Cr_4O_{16}$ (c) and $K_2V_2Cr_6O_{16}$ (d).

Finally, we investigated the crystal structure of $K_2V_2Cr_6O_{16}$. This compound consists of one each of K and V, three Cr, four each of O1 and O2 atoms [see figure 1(d)]. In $K_2V_2Cr_6O_{16}$, RFCCs are also formed by the Cr, V and O2 atoms and found in the ab plane. In each RFFC, the single V and one Cr atom at one opposite diagonal positions in conjugation with two O2 atoms are positioned at the z=0 or z=1 plane. On the other hand, other opposite diagonal positions contain two Cr atoms. These two Cr atoms and two O2 atoms are positioned at the z=1/2 plane. Nevertheless, Cr and V atoms at z=0 or z=1 are shifted from the mirror plane [see figure 2(d)]. Consequently, Cr or V dimers form in the Cr-Cr or V-V double chains, which further facilitates Peierls distortion within the crystal. Indeed, two Cr atoms at z=1/2 plane preserve every mirror symmetry [see figure 2(d)], which results in uniform Cr-Cr distances in the double chains. Therefore, mirror symmetry is practically lost in 25% and 75% Cr-doping for V only. However, structural optimizations were not carried out in the present study, which will be corroborated in our future work.

3.2. Electronic and magnetic properties

We first concentrated on the spin-polarized electronic band structure calculations of pure $K_2V_8O_{16}$ at HT structure. It is found that the system is metallic in the spin majority channel [see figure 3(a)], while insulating in the spin minority channel (not shown in figure 3). Considering both spin channels, the system is a half-metal. The formal valence of V in the present system is +3.75 i.e. the available V-3d electrons is 1.25.

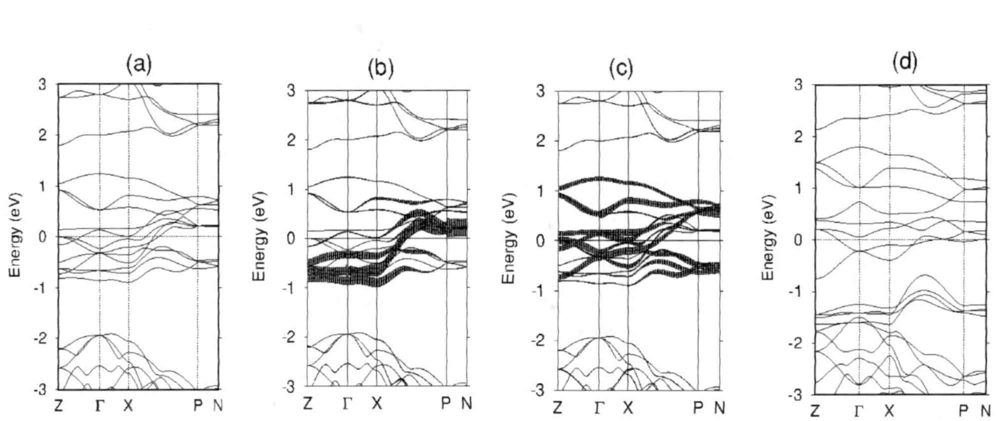

Figure 3. The LDA spin-polarized electronic band structures of $K_2V_8O_{16}$ for the spin majority channel (a). The d_{xy}, $d_{yz/xz}$ states are illustrated in (b) and (c) respectively. The band structure for the spin majority channel using U=4 eV is illustrated in (d). The Fermi level is set at zero in the energy scale.

These electrons are found in V-t_{2g} bands, while e_g bands are found unoccupied (not shown in figure 3). The partial band structures of V-t_{2g} states are illustrated in figure 3. It is obvious from this figure that the available 1.25 electrons are shared by V-d_{xy}, d_{yz} and d_{xz}states. Therefore, none of the V-t_{2g} states are completely filled by the available 3d electrons. More interestingly, d_{yz} and d_{xz} states are exactly degenerate [see figure 3(c)]. The partially occupied t_{2g} states results in metallic behaviour of $K_2V_8O_{16}$. Accordingly, an orbital fluctuation between d_{xy} and d_{yz} states is also observed due to the delocalization of electrons in these states. The band structure of the present system using U=4 eV is displayed in figure 3(d). The scenario does not improve, i.e. the system remains in its half-metallic phase. We found that the occupied states become more occupied and unoccupied states become more unoccupied under the influence of U (not shown in figure 3). Nevertheless, no band gap opens upon the application of U=4 eV.

Next, band dispersion calculations of $K_2V_6Cr_2O_{16}$ (25% Cr-doping) were carried out in the LDA treatment. The electronic band structures of this compound for the spin majority channel is shown in figure 4(a). The system remains metallic at 25% Cr doping. The partial band structures of Cr/V-t_{2g} states are also displayed in figure 4. The available electrons in the Cr and V-t_{2g} states are 2.25 and 1.25 respectively. Out of 2.25 electrons, one completely occupies the Cr-d_{xy} state [see figure 4(b)], whereasd_{yz} and d_{xz} states have an equal probability to be occupied by the remaining 1.25 electrons. Thus, these two states are exactly degenerate. Furthermore, all V-t_{2g} states share the available 1.25 electrons. Indeed, occupancy of d_{xy} states increases significantly [figure 4(d)], whereas that decreases markedly for $d_{yz/xz}$ states [see figure 4(e)].

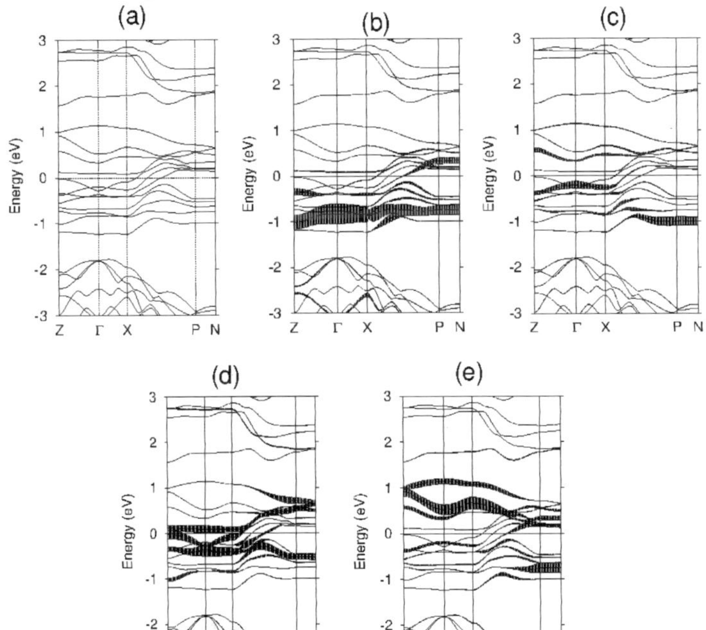

Figure 4. The LDA spin-polarized electronic band structures of $K_2V_6Cr_2O_{16}$ for the spin majority channel (a). The Cr-d_{xy}, $d_{yz/xz}$ states are illustrated in (b), (c) and V-d_{xy}, $d_{yz/xz}$ states are illustrated in (d), (e). The Fermi level is set at zero in the energy scale.

Nevertheless, the electrons in the degenerate states are delocalized near E_F. These delocalized d electrons are responsible for orbital fluctuations between the partially filled Cr-$d_{yz/xz}$andV-t_{2g} states near the close vicinity of E_F. This sharing of electrons by the partially filled Cr/V-t_{2g} states is further responsible for the metallic behaviour of $K_2V_6Cr_2O_{16}$. Besides, double exchange interaction between electrons of the partially occupied Cr/V-t_{2g} states facilitates FM in this material. The e_g states were observed unoccupied and shifted well above E_F (not shown in figure 4). This material is also insulating in the spin minority channel (not shown in figure 4). Therefore, considering two spin channels, the system is half-metal. More interestingly, Peierls distortion due to the formation of dimers and symmetry lowering is still not capable to open a gap near E_F. Figure 5 displays the electronic band structure of the present material for the spin majority channel using U=4 eV. The system encounters MIT at U=4 eV with a band gap of 0.36 eV. Noteworthy, the bands below -1.0 eV are derived from O-2p states which are hybridized with Cr/V-t_{2g} states. To elucidate the origin of MIT in $K_2V_6Cr_2O_{16}$, we further calculated the partial band structures of Cr/V-t_{2g} states as illustrated in figures 5(b)-(d). It is unambiguous from this figure that all the Cr-d_{xy} figure (b), d_{yz} and d_{xz} figure (c) states are occupied by one electron each. The d_{yz} and d_{xz} states are also degenerate in this case. Indeed, the d_{xy} state is more occupied compared to $d_{yz/xz}$ states.

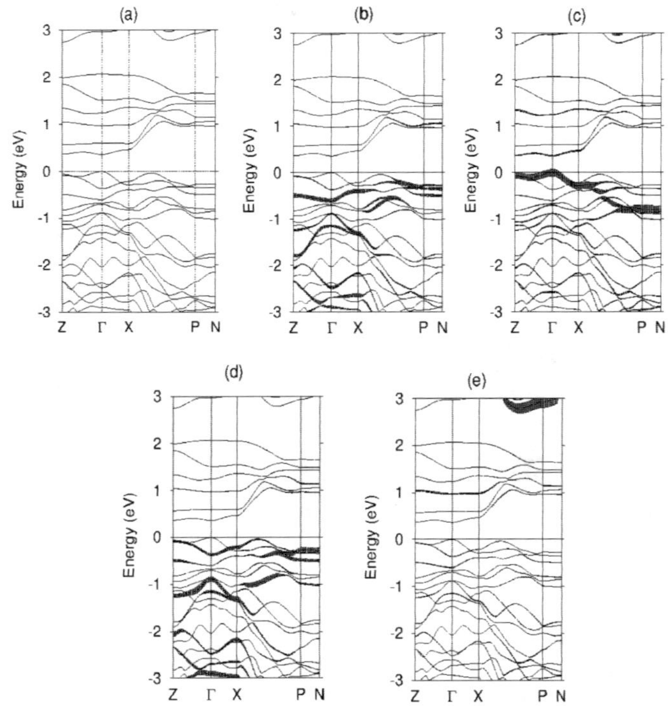

Figure 5. The LDA+U (U= 4 eV) electronic band structures of $K_2V_6Cr_2O_{16}$ for the spin majority channel (a). The partial band structures of Cr-d_{xy}, Cr-$d_{yz/xz}$,V-d_{xy} and V-$d_{yz/xz}$ states are illustrated in (b), (c), (d) and (e) respectively. The Fermi level is set at zero in the energy scale.

Noteworthy, ⅓ of the V-t_{2g} sates becomes filled. We found that the V-d_{xy} state is occupied [see figure 5(d)], while d_{yz} and d_{xz} states become unoccupied [see figure 5(e)]. Therefore, out of six available electrons per formula unit (KV_3CrO_8), three occupy three Cr-t_{2g} states and remaining three electrons occupy three V-d_{xy} states**.** It is also observable that O-2p states are hybridized with all the Cr-t_{2g} and V-d_{xy} states, which further stabilize FM in the insulating $K_2V_6Cr_2O_{16}$. Noteworthy, p-d hybridizations for Cr/V-d_{xy} states are much higher in comparison with that for Cr-$d_{yz/xz}$ states. The inclusion of electronic correlations among the degenerate Cr/V-3d states is responsible for reduction in the orbital fluctuations. About 0.25 electron from each of V atoms transfers to the Cr-t_{2g} states such that they become fully occupied. Due to this charge (electron) transfer from V to Cr-t_{2g} states, the number of occupied 3d electrons of Cr and V becomes three (Cr^{3+}) and one (V^{4+}) respectively. As a result, charge orderings in the fashion of Cr^{3+}-V^{4+} and V^{4+}-V^{4+} occur, which results in the opening of the band gap when U is included. In particular, each of Cr-d_{yz} and d_{xz} state is now occupied by one electron each. Consequently, all the three Cr-t_{2g} states and only V-d_{xy} state become occupied and are shifted below E_F, whereas all the Cr/V-e_g states remain unoccupied (not shown in figure 5). As a

IOP Conf. Series: Materials Science and Engineering 1183 (2021) 012004 doi:10.1088/1757-899X/1183/1/012004

result, a band gap opens near E_F and the system thereby encounters MIT. Therefore, Peierls like structural distortions in conjugation with charge ordering and electrons correlations are accounted for the MIT in $K_2V_6Cr_2O_{16}$. Band gaps observed for U=4, 5 and 6 eV are 0.36, 0.59 and 0.79 eV respectively. Occupied Cr-t_{2g} and V-d_{xy} states are directed so much below E_F that they are strongly hybridized with O-2p states. Since Cr-3d O-2p hybridization increases significantly, the occupancy of Cr-t_{2g} bands increases due to band filling effect. Therefore, the insulating phase of $K_2V_6Cr_2O_{16}$ is accompanied by this band filling effect driven by breaking in symmetry. We also observed that Cr-3d O-2p coupling increase significantly due to the application of U, which is further responsible for the ferromagnetism in insulating $K_2V_6Cr_2O_{16}$.

In order to complete the study, we investigated the electronic and magnetic properties of $K_2V_4Cr_4O_{16}$ (50% Cr doping). The band structure for the spin majority channel is shown in figure 6(a). The system is metallic in this channel. However, no p-d hybridization is observed in this case. The partial band structures of Cr/V-t_{2g} states are also exhibited in figures 6(b)-(d). It is noticeable that Cr-d_{xy} [figure 6(b)], $d_{yz/xz}$ states [figure 6(c)] are occupied. Besides, V-d_{xy} state [see figure 6(d) is found partially occupied, whereas $d_{yz/xz}$ states are found unoccupied (not shown in figure 6). It should be noted that the total number of d electrons per f.u. ($KV_2Cr_2O_8$) is seven. Now simple electron counting tells us that V-t_{2g} states will be partially occupied which is reflected in the partial band structures of V-t_{2g} sates. Out of which, six electrons occupy six t_{2g} states of two Cr atoms and the remaining electron is equally shared by two d_{xy} states of two V atoms. In this sense no charge ordering occurs due to transfer of charge in $K_2V_4Cr_4O_{16}$. Since the number of electrons increases due to Cr doping, electron correlation becomes more important for the FM of the present material. Noteworthy, application of U =4 eV is still not able to open a gap at E_F [see figure 6(e)] due to absence of charge ordering. However, the occupied levels become more occupied and shifted far below E_F, whereas the unoccupied levels become more unoccupied and shifted far above E_F under the influence of U=4 eV, which is also clearly visible from figure 6(e). Furthermore, due to high symmetry of the crystal, Cr-$d_{xy/xz}$ states remain degenerate and thereby none of them become completely occupied. It is worth mentioning that the p-d hybridization increases significantly [figure 6(e)] due application of U. Besides, Cr-O coupling again is responsible for FM of the present material.

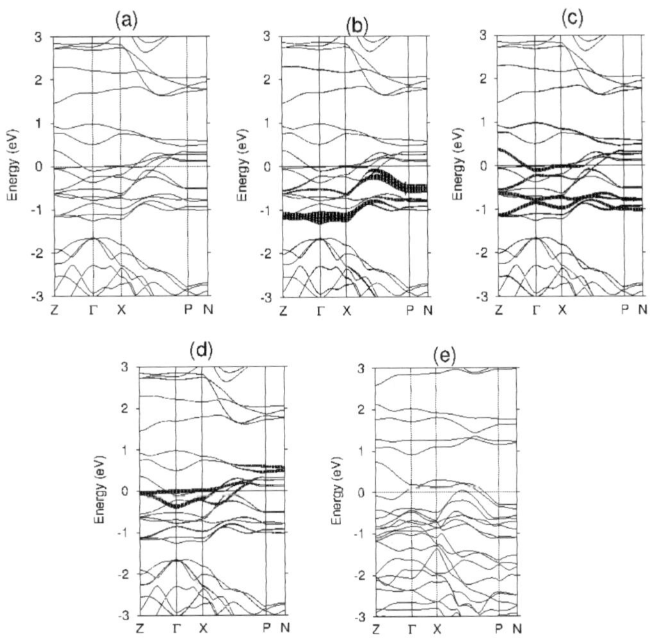

Figure 6. LDA spin-polarized band structure of $K_2V_4Cr_4O_{16}$ (a) for the spin majority channel. The partial band structure of Cr-d_{xy}, $d_{yz/xz}$ and V-d_{xy} states are depicted in (b), (c) and (d) respectively. The band structure of this material for U =4 eV is also illustrated in (e).

Finally, we concentrated on the electronic and magnetic properties of $K_2V_2Cr_6O_{16}$ (75% Cr doping). The band structure for the spin majority channel is shown in figure 7(a). It is evident from this figure that the system remains metallic. The partial band structures of Cr-d_{xy}, $d_{yz/xz}$ and V-d_{yz} states are illustrated respectively in figures 7(b), (c) and (d). It is obvious that Cr-d_{xy} states are fully occupied, whereas Cr-$d_{yz/xz}$ states are almost (83%) occupied. It is also evident that V-d_{xy} state is mostly unoccupied. We also found that V-$d_{yz/xz}$ states become completely unoccupied (not shown in figure 7). Consequently, the number of electrons in Cr and V-t_{2g} states become 2.67 and almost zero i.e. $Cr^{3.33+}$-V^{5+} charge ordering occurs. We have also applied of U =4 eV, but no gap opens at E_F [see figure 7(e)], because transferred electrons are not sufficient to completely fill Cr-$d_{yz/xz}$ states. Although the cooperative effect of Peierls instability due to structural distortion and Coulomb interactions is acted upon $K_2V_2Cr_6O_{16}$, it remains metallic due to lack of sufficient electrons to completely occupy Cr-$d_{yz/xz}$ states. Furthermore, p-d hybridization increases markedly as can be seen from figure 7(e). We observed that Cr-O coupling also augmented significantly, which is accounted for the FM of $K_2V_2Cr_6O_{16}$.

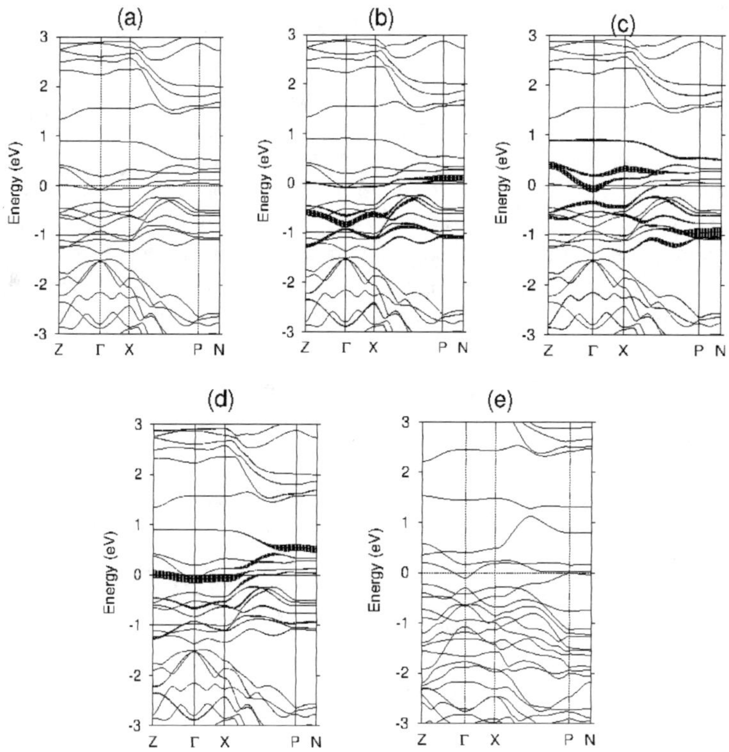

Figure 7. LDA band structure of $K_2V_2Cr_6O_{16}$ (a) for the spin majority channel. The partial band structure of Cr-d_{xy}, $d_{yz/xz}$ and V-d_{xy} states are depicted in (b), (c) and (d) respectively. The band structure of this material for U =4 eV is also illustrated (e).

The total ground states energies (E_{tot}) of pure and different Cr-doped $K_2V_8O_{16}$ as calculated in the PM, AFM and FM configuration are reported in Table 1. It is unambiguous that E_{tot} is the minimum for FM calculations for all three doping levels. This result indicates that ground state of $K_2V_8O_{16}$, $K_2V_6Cr_2O_{16}$, $K_2V_4Cr_4O_{16}$ and $K_2V_2Cr_6O_{16}$ are ferromagnetic. Furthermore, E_{tot} increases significantly with Cr concentration for all three calculations. It should be noted that E_{tot} enhances significantly about 203, 406 and 608 eV respectively for 25%, 50% and 75% Cr-doping in $K_2V_8O_{16}$. Therefore, strength of FM is augmented significantly with Cr-concentration in $K_2V_8O_{16}$.

Table 1. Total ground state energies (eV) for $K_2V_8O_{16}$, $K_2V_6Cr_2O_{16}$, $K_2V_4Cr_4O_{16}$, $K_2V_2Cr_6O_{16}$ and in the PM, AFM and FM configurations.

Compounds	$K_2V_8O_{16}$	$K_2V_6Cr_2O_{16}$	$K_2V_4Cr_4O_{16}$	$K_2V_2Cr_6O_{16}$
PM	-9985.52	-10188.31	-10391.11	-10593.90
AFM	-9985.53	-10188.35	-10391.18	-10593.97
FM	-9985.55	-10188.38	-10391.22	-10594.0

4. Conclusion

The Presence of Cr at the B-site renders the hollandite $A_2B_8O_{16}$ ferromagnetism but that of V at B-site enhances the T_{MI}. We investigated the effect of Cr doping at the vanadium site in a systematic way, *i. e.* 25%, 50% and 75% doping. It was found that although the first and third led to considerable break in mirror symmetry in the square well but only in the first case, an energy gap opens up when Hubbard-like U is applied. Hence we find the necessary ingredient of obtaining MIT in hollandites is to Peierls instability (dimerization)-driven band filling in the transition metal atoms and Coulomb correlations. It is also explored that strength of FM increases with Cr doping.

5. Acknowledgements

The authors thank the West Bengal State University, Kolkata-700126 and Department of Higher Education, Government of West Bengal, India for financial and infrastructure support to carry out this work.

6. References

[1] Tamada, O., Yamamoto, N., Mori, T. and Endo, T., 1996. The crystal structure of K2Cr8O16: The hollandite-type full cationic occupation in a tunnel. *Journal of Solid State Chemistry*, *126*(1), pp.1-6.

[2] Hasegawa, K., Isobe, M., Yamauchi, T., Ueda, H., Yamaura, J.I., Gotou, H., Yagi, T., Sato, H. and Ueda, Y., 2009. Discovery of ferromagnetic-half-metal–to–insulator transition in K 2 Cr 8 O 16. *Physical Review Letters*, *103*(14), p.146403.

[3] Sakamaki, M., Konishi, T. and Ohta, Y., 2009. K 2 Cr 8 O 16 predicted as a half-metallic ferromagnet: Scenario for a metal-insulator transition. *Physical Review B*, *80*(2), p.024416.

[4] Khomskii, D.I. and Sawatzky, G.A., 1997. Interplay between spin, charge and orbital degrees of freedom in magnetic oxides. *Solid State Communications*, *102*(2-3), pp.87-99.

[5] Mahadevan, P., Kumar, A., Choudhury, D. and Sarma, D.D., 2010. Charge ordering induced ferromagnetic insulator: K 2 Cr 8 O 16. *Physical Review Letters*, *104*(25), p.256401.

[6] Toriyama, T., Nakao, A., Yamaki, Y., Nakao, H., Murakami, Y., Hasegawa, K., Isobe, M., Ueda, Y., Ushakov, A.V., Khomskii, D.I. and Streltsov, S.V., 2011. Peierls mechanism of the metal-insulator transition in ferromagnetic hollandite K 2 Cr 8 O 16. *Physical Review Letters*, *107*(26), p.266402.

[7] Nakao, A., Yamaki, Y., Nakao, H., Murakami, Y., Hasegawa, K., Isobe, M. and Ueda, Y., 2012. Observation of structural change in the novel ferromagnetic metal–insulator transition of K2Cr8O16. *Journal of the Physical Society of Japan*, *81*(5), p.054710.

[8] Isobe, M., Koishi, S., Yamazaki, S., Yamaura, J.I., Gotou, H., Yagi, T. and Ueda, Y., 2009. Substitution Effect on Metal–Insulator Transition of K2V8O16. *Journal of the Physical Society of Japan*, *78*(11), pp.114713-114713.

[9] Isobe, M., Koishi, S. and Ueda, Y., 2007. Ti-doping effect on the MI transition of hollandite vanadate, K2V8O16. *Journal of magnetism and magnetic materials*, *310*(2), pp.888-889.

[10] Isobe, M., Koishi, S., Kouno, N., Yamaura, J.I., Yamauchi, T., Ueda, H., Gotou, H., Yagi, T. and Ueda, Y., 2006. Observation of metal–insulator transition in hollandite vanadate, K2V8O16. *Journal of the Physical Society of Japan*, *75*(7), p.073801.

[11] Waki, T., Kato, H., Kato, M. and Yoshimura, K., 2004. Metal–Insulator Transition in Bi x V8O16: 51V NMR Study. *Journal of the Physical Society of Japan*, *73*(1), pp.275-279.

[12] Kuwabara, T., Isobe, M., Gotou, H., Yagi, T., Nishio-Hamane, D. and Ueda, Y., 2012. Synthesis, Structure, and Electromagnetic Properties of Manganese Hollandite, K x Mn8O16. *Journal of the Physical Society of Japan*, *81*(10), p.104701.

[13] Sugiyama, J., Ikedo, Y., Mansson, M., Ofer, O., Ansaldo, E.J., Brewer, J.H., Chow, K.H., Lord, J., Ohta, H., Michioka, C. and Yoshimura, K., 2012. Magnetic nature of water intercalated Na0. 35CoO2. *Physics Procedia*, *30*, pp.266-270.

[14] Isobe, M., Koishi, S., Kouno, N., Yamaura, J.I., Yamauchi, T., Ueda, H., Gotou, H., Yagi, T. and Ueda, Y., 2006. Observation of metal–insulator transition in hollandite vanadate, K2V8O16. *Journal of the Physical Society of Japan*, *75*(7), p.073801.

[15] Horiuchi, S., Shirakawa, T. and Ohta, Y., 2008. Theoretical study of the electronic states of hollandite vanadate K 2 V 8 O 16. *Physical Review B*, *77*(15), p.155120.

[16] Komarek, A.C., Isobe, M., Hemberger, J., Meier, D., Lorenz, T., Trots, D., Cervellino, A., Fernández-Díaz, M.T., Ueda, Y. and Braden, M., 2011. Dimerization and Charge Order in Hollandite K 2 V 8 O 16. *Physical review letters*, *107*(2), p.027201.

[17] Toriyama, T., Konishi, T. and Ohta, Y., 2012. Mechanism of the metal-insulator transition of hollandite vanadate K2V8O16. In *Journal of Physics: Conference Series* (Vol. 400, No. 3, p. 032104). IOP Publishing.

[18] Ishige, Y., Sudayama, T., Wakisaka, Y., Mizokawa, T., Wadati, H., Sawatzky, G.A., Regier, T.Z., Isobe, M. and Ueda, Y., 2011. Interplay between Mott physics and Peierls physics in hollandite-type vanadates with a metal-insulator transition. *Physical Review B*, *83*(12), p.125112.

[19] Kim, S., Kim, B.H., Kim, K. and Min, B.I., 2016. Metal-insulator transition in a spin-orbital-lattice coupled Mott system: K 2 V 8 O 16. *Physical Review B*, *93*(4), p.045106.

[20] Fukuyama, H., Ohta, Y., Shirakawa, T. and Horiuchi, S., 2008. Superconductivity in B-doped diamonds. *Journal of Physics and Chemistry of Solids*, *69*(12), pp.3265-3268.

[21] Muraoka, Y., Noami, K., Wakita, T., Hirai, M., Yokoya, T., Kayo, Y., Muro, T. and Tamenori, Y., 2011. Synthesis and physical properties of the hollandite-type titanium oxide KxTi8O16. *Physica Status Solidi C*, *8*(2), pp.555-557.

[22] Hohenberg, P. and Kohn, W., 1964. Inhomogeneous electron gas. *Physical review*, *136*(3B), p.B864.

[23] Anderson, O.K., 1974. Linear methods in band theory. Physical Review B 12, p. 3060.

[24] Anisimov, V.I., Aryasetiawan, F. and Lichtenstein, A.I., 1997. First-principles calculations of the electronic structure and spectra of strongly correlated systems: the LDA+ U method. *Journal of Physics: Condensed Matter*, *9*(4), p.767.

[25] Abriel, W., Rau, F. and Range, K.J., 1979. New compounds A2− xV8O16 (A= K, Tl) with hollandite type structure. *Materials Research Bulletin*, *14*(11), pp.1463-1468.

Investigation and Analysis of Renewable Energy from Bio and Solid Waste

M. Sakthivel[1*], A. Arun[1], E. Akram Hussainali[2], G. Balamurugan[2]

[1]Associate Professor, Department of Mechanical Engineering, IFET College of Engineering, Villupuram, Tamilnadu, India.
Department of Mechanical Engineering, IFET College of Engineering, Villupuram, Tamilnadu, India.

[*]Corresponding author Email: sakthivel@ifet.ac.in

Abstract. This investigation explores the possibility of utilizing bio and solid waste to produce power as renewable energy in every residential house. Fossil fuels are depleting very fast and environmental pollution is also caused due to their combustion. An effort is taken to generate electric power by the high combustion rate methane gas released by the bio and solid wastes from the residence. Anaerobic digestion is adopted to produce methane gas from the waste dumped in any house. This methane gas is made to run a single cylinder diesel engine coupled to a single phase alternator. Power thus generated will be sufficient to any single residential house.

Keywords: *Renewable Energy; Bio and Solid Waste; Anaerobic Digestion; Methane Gas; Diesel Engine; Alternator*

1. Introduction

Rapid depletion of coal, natural gas and petroleum fossil fuels are taking place due to the high demand in energy faced by the world. Ozone layer depletion, greenhouse effect and acid rains are caused globally due to the combustion of fossil fuels. The pollution to the environment thus created poses great difficulty to the total life on the planet. Only if a system of cleaner energy is replacing the existing system of fossil fuels, the problems faced globally can be solved. Ozone layer depletion, greenhouse effect, acid rains and pollution will be eliminated if a clean and efficient fuel like methane is used for combustion. Energy produced from the renewable sources like solar, wind and methane results in a permanent energy system.

Non-renewable energy system of fossil fuels emits pollutants which cause greater damage than the renewable energy system of methane. Efforts on long-term research are to be taken using solar energy and water to economically produce methane. Worldwide CO_2 emission is to be lowered to reduce the greenhouse effect and the climate change. For lowering the emissions of CO_2 and reducing various environmental pollutions, use of methane is an alternative on long-term, even though it is not economically viable at present. Fossil fuel to methane conversion can solve most of the problems. Best way is to produce methane from pollution free sources.

Content from this work may be used under the terms of the Creative Commons Attribution 3.0 licence. Any further distribution of this work must maintain attribution to the author(s) and the title of the work, journal citation and DOI.
Published under licence by IOP Publishing Ltd

2. MATERIALS AND METHODS

2.1. Bio Mass Gas Preparation

Bio and solid wastes generate more power and the cost is also cheaper compared to power generation from the cow dung. Anaerobic digestion and Aerobic digestion are two methods for bio gas production. In Anaerobic digestion, methane is released along with water and carbon-dioxide by the conversion of biodegradable organic matters without using oxygen. The anaerobic process reduces and inactivates pathogens.

$$\text{Organic matter} + H_2O \rightarrow CH_2 + CO_2 + H_2O$$

Anaerobic digestion reduces volatile solids to 60 per cent by oxidation of biodegradable and microbial cellular matter with a major mass reduction in sludge by generating and stabilizing cell mass.

The action of heterotrophic bacteria in aerobic digestion hydrolyses and converts biodegradable particulate organic matter into water, carbon dioxide and active biomass.Nitrate-N or dissolved oxygen is used for oxidation as terminal electron acceptor. Raw sludge properties control the aerobic digestion progress. Aerobic digestion process is costly and energy recovery is also nil. Because of above facts, anaerobic digestion process is preferred to the production of methane gas for power generation.

2.2. CHEMICAL PROCESS

The following chemical processes are normally adopted for converting natural gas to methane:

- MSR process (Methane steam reforming)
- POX process (Partial methane oxidation)
- ATR process or Casale's process (Auto thermal reforming)

MSR process is a mature technology, cost effective and most common method. Mass and heat transfer issues are the significant disadvantages. Hydrogen and methane conversion take place due to endothermic reaction at a temperature of 850°C and at a pressure of 2.5 MPa with nickel as catalyst.

Stage 1. $H_2O \text{ (steam)} + CH_4 \rightarrow 3H_2 + CO \text{ (endothermic)}$ $\Delta H = 206 \text{ kJ}$

Stage 2. $H_2O \text{ (steam)} + CO \rightarrow H_2 + CO_2$ (exothermic) $\Delta H = -41 \text{ kJ}$

Heat needed in stage 1 is got from surplus heat in stage 2 and combustion of methane feedstock.

POX process is a mild exothermic process without external heating and thereby a compact design for the conversion of methane. Only a limited amount of oxygen reacts with methane. In the final product, other compounds and small amount of carbon di oxide exist.

$$CH_4 + O_2 \rightarrow CO + 2H_2 \text{ (exothermic)} \qquad \Delta H = -36 \text{ kJ}$$

IOP Conf. Series: Materials Science and Engineering 1183 (2021) 012005 doi:10.1088/1757-899X/1183/1/012005

Figure 1. Prototype model biogas preparation

ATR process is the manifestation of both the reactions of MSR and POX. In general, ATR operates at a temperature between 950 to 1050°C and a pressure of 30 to 50 bars. $H_2$is injected along with steam to produce CH_4. Efficiency is reduced due to purification of output gas and large investment is also needed. Hence Partial oxidation with anaerobic digestion is adopted for the production of methane gas.

In Proto type model shown in Fig.1, three barrels of each 200 lit.Capacity is used for the biogas preparation from the bio and solid wastes generated from the residential house. For every 25 lit. of waste 1 cft. Methane gas is produced which is sufficient to light a 25W bulb for 6 hours.

2.3. DIESEL ENGINE AND ALTERNATOR

2.3.1 Engine Specification

Type	: Kirloskar AVI 4-stroke water cooled
Cylinder	: single
BHP	: 5HP
Speed	: 1500 rpm
Fuel	: Dual Fuel (Diesel/Bio gas)
Bore	: 87.5mm
Stroke length	: 110mm
Compression ratio	: 16:1

2.3.2 Alternator

Type	: Single phase
Range	: 1.5 KVA
Speed	: 1500 rpm

3. RESULTS AND DISCUSSION

3.1 Performance trails and results

Kirloskar engine with dual fuel is used for the performance trails. A single phase alternator is coupled to the engine. Resistive loads are applied during the trails. At various loads trails are conducted and the performance output results are tabulated as shown in the tables 1, 2 and 3. And Performance analysis results are shown in figures 2 to 8.

Table 1. Total fuel consumption and efficiencies at various kw brake power

S.NO	BRAKE POWER kW	TFC Kg/Hr		BTE %		ITE %		ME %	
		Diesel	Dual mode	Diesel	Dual mode	Diesel	Dual mode	Diesel	Dual mode
1	1.0	0.65	0.47	13.2	14.1	28	23	38	53
2	2.0	0.71	0.51	23.0	26.0	38	35	62	83
3	3.0	0.89	0.59	28.0	32.0	39	37.5	68	82
4	3.5	0.95	0.66	26.0	30.0	38	34.2	67	81

Table 2. Fuel consumption at various kW brake power

S.NO	BRAKE POWER kW	FUEL CONSUMPTION %	
		Diesel	Biogas
1	1.606	84	16
2	2.120	81	19
3	2.620	77	23
4	3.520	75	25
	Average	69	21

Table 3. Exhaust emission results in ppm

S.NO	FUEL	EMISSION (ppm)			
		CO	CO_2	NO	CH_4
1	DIESEL	70	107	50	90
2	DIESEL AND BIO GAS	57	37	41	85

3.2 Total Fuel Consumption Vs Brake Power

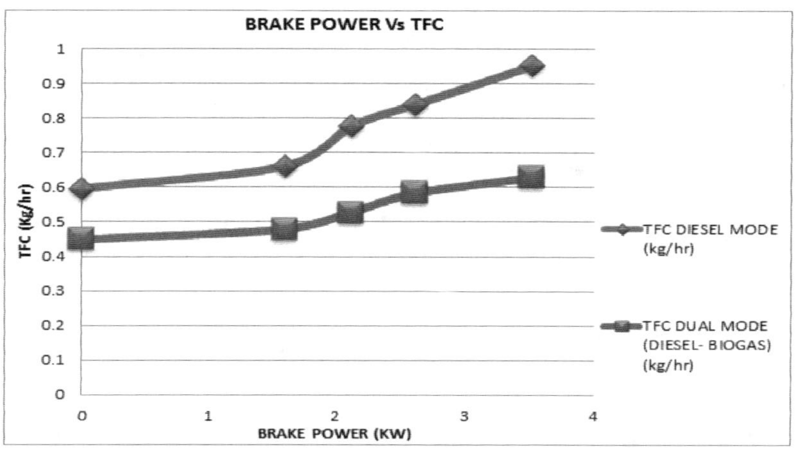

Figure 2. Total Fuel Consumption Vs Brake Power

3.3 Brake Thermal Efficiency Vs Brake Power

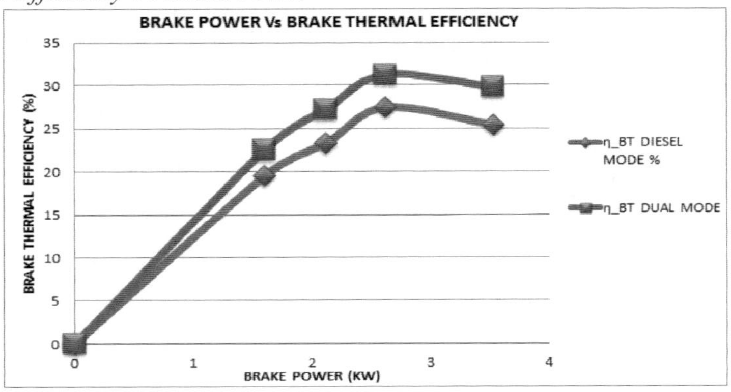

Figure 3. Brake Thermal Efficiency vs Brake Power

3.4 Indicated Thermal Efficiency Vs Brake Power

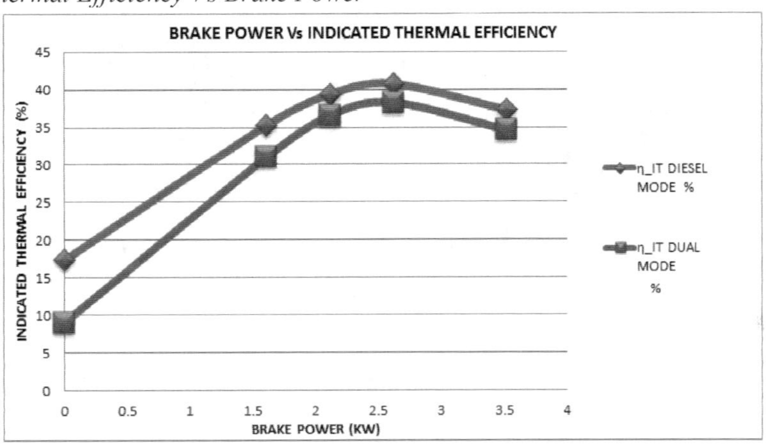

Figure 4. Indicated Thermal Efficiency Vs Brake Power

3.5 Mechanical Efficiency Vs Brake Power

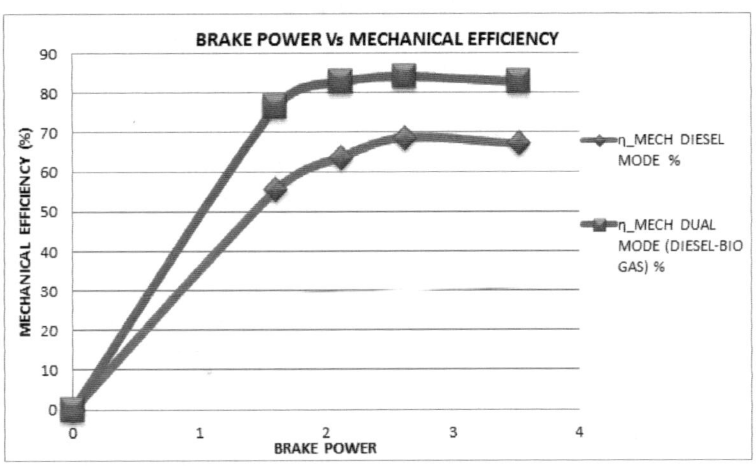

Figure 5. Mechanical Efficiency Vs Brake Power

3.6Fuel Consumption for Dual Mode at Various Load Condition

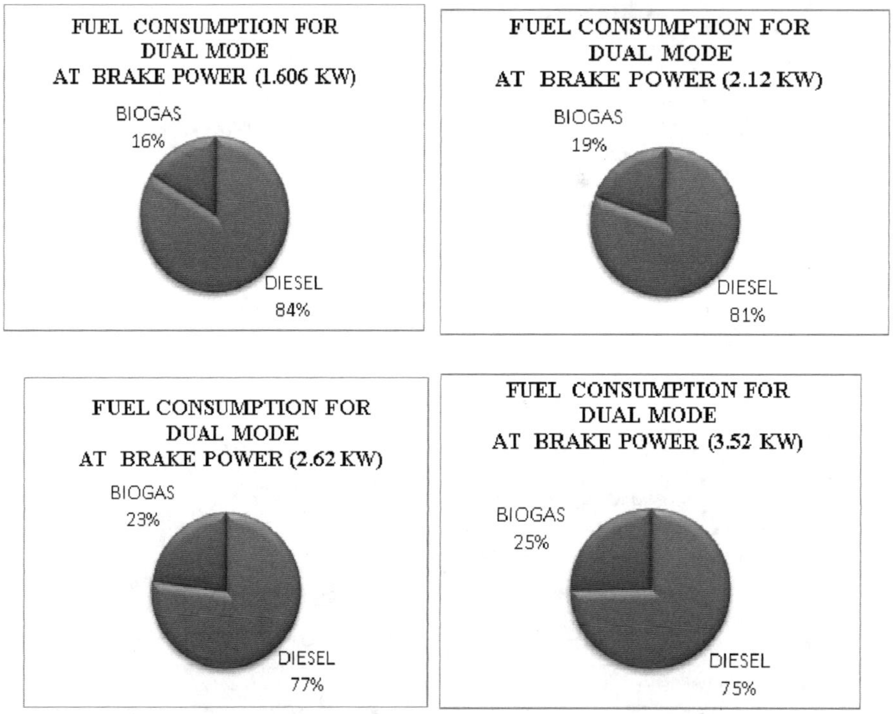

Figure 6. Load Condition Vs Fuel Consumption

3.7.Average Fuel Consumption for Dual Mode (Diesel – Biogas)

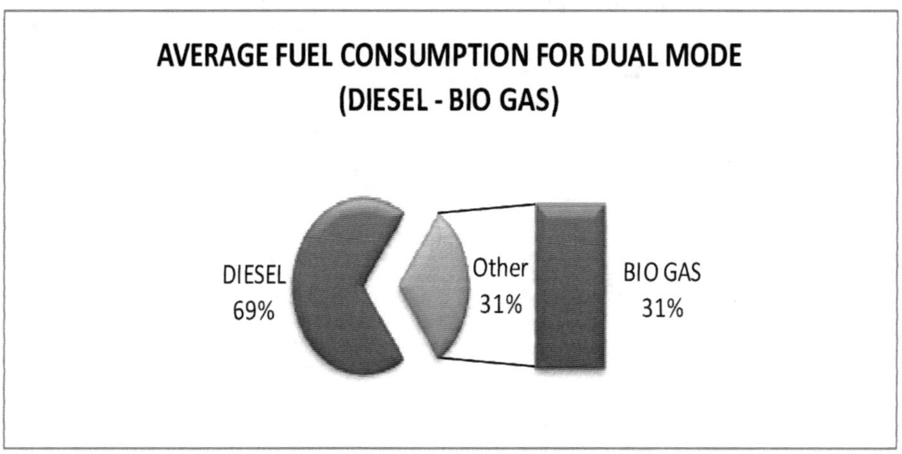

Figure 7. Average fuel consumption

It is noticed from the analysis as shown in fig.7 that the diesel consumption is 69% and 31% in the dual mode.

3.8 Emission Result

Engine Load: Ideal Load **Room Temperature:** 25±5°C

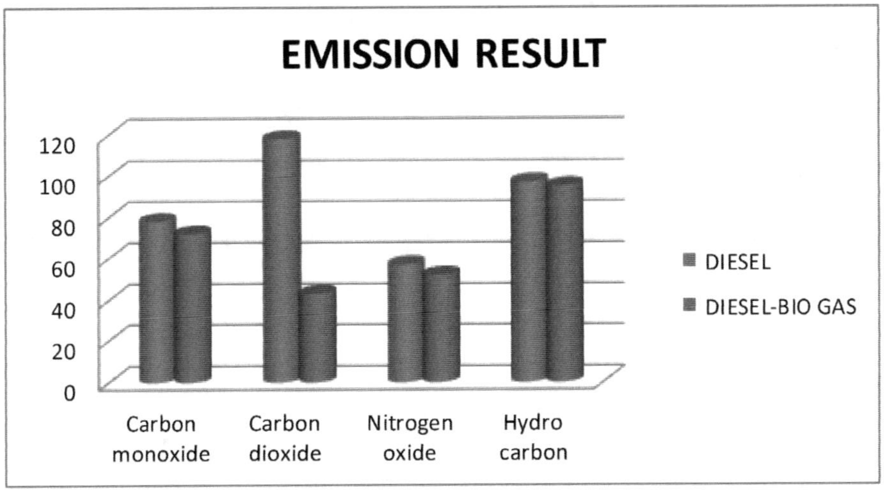

Figure 8. Emission

In the dual mode emission is also found to get reduced as shown in Fig.8.

4. SUMMARY AND CONCLUSION

The investigation and analysis of renewable energy from bio and solid waste observation reveals that smaller power generation set ups are possible and feasible in every residence. In a simpler manner and at a lower cost, the daily demand of every residence can be met. The production of renewable energy by this way, waste disposal also becomes easier. During analysis, power loss due to friction in dual mode is found to get reduced by 23% along with a reduction of 21% in diesel consumption. Mechanical efficiency is found to increase by around 16% on average in dual mode as compared to diesel mode. Similarly, emission of carbon monoxide, carbon dioxide, nitrogen oxide and hydro carbon are found to get reduced from 70 to 57 ppm, 107 to 37 ppm, 50 to 41 ppm and 90 to 85 ppm respectively from diesel mode to dual mode. This investigation and analysis clearly shows that the renewable energy from bio and solid waste is more beneficial.

5. REFERENCES

[1] N.H.S.Ray, "A Study on Application of Biogas as fuel in Compression Ignition Engines", Dept. of Mech. Engg. GEC, BijuPatnaik University of Technology, Odisha, India, Vol. 3 Issue 1 October 2013.

[2] M. Nithin, Choppara. Rishikanth, R.Vijayan, M. Sakthivel, "Performance Test on Single Cylinder IC Engine by Using Cow Dunk as a Bio gas", International Journal of Emerging Technology and Advanced Engineering, Volume 6, Issue 3, March 2016.

[3] Angelidaki, I., Ellegaard, L., Ahring, B.K., 1993 "A mathematical model for dynamic simulation of anaerobic digestion of complex substrates: focusing on ammonia inhibition", Biotechnology Bioengineering, 42: 159 - 166.

[4] Rittman, B.E., "Opportunities for renewable bioenergy using microorganisms", 2008, Biotechnology and Bioengineering, 100: 203-212.

IOP Conf. Series: Materials Science and Engineering 1183 (2021) 012005 doi:10.1088/1757-899X/1183/1/012005

[5] Prithivirajan.K, Prakash.E, Sakthivel.M., "The Performance of IC Engine with (Diesel-Hydrogen) Dual Fuel", International Journal of Advances in Mechanical and Civil Engineering, ISSN: 2394-2827 Volume-2, Issue-3, June-2015.

[6] Far land, M. J. M., 2000, Bio solids engineering, McGraw Hill Professional, London, UK.

[7] Rao, P. V., Baral, S. S., Ranjan, D., Srikanth, M., 2010, "Biogas generation potential by anaerobic digestion for sustainable energy development in India", Renewable and Sustainable Energy Reviews, 14: 2086 - 2094.

[8] Wieland, P., 2010, "Biogas production: current state and perspectives", Applied Microbiology and Biotechnology, 85: 849 - 860.

[9] Bagi, Z., ACS, N., Balint, B., Horvath, L., Dobo, K., Perei, K.R., Rakhely, G., Kovacs, K.L., "Biotechnological intensification of biogas production", Applied Microbiology and Biotechnology, 76: 473 - 482.

[10] Schink, B., 1997, "Energetics of syntrophic cooperation in methanogenic degradation", Microbiology and Molecular Biology Review, 61:262 - 280.

[11] Ferrer, I., Ponsa, S., Vazquez, F., Font, X., 2008 "Increasing biogas production by thermal (700 C) sludge pre-treatment prior to thermophilic anaerobic digestion", Biochemical Engineering Journal, 42: 186 - 192.

[12] Calvin A. Cherrya, Michael Riosa, AleiaMcCordb, Sarah Stefanosb, GiriVenkataramanana "Portable Electrification using Biogas Systems" procedia engineering 78(2014) page 317-326.

[13] Jagadish H Patila, Mal Antonyraja, Shankar Bb B, Mahesh Kumar Shettyc, B P PradeepKumard "Anaerobic co-digestion of water hyacinth and sheep waste" energy procedia 52(2014) page from 572-578.

[14] SiripornYossan, Li Xiao, PoonsukPrasertsan, "International Journal of Methane Energy", Zhen He.6 August 2013. Volume 38, Issue 23, Pages 9619-9624

[15] Andrew Whiting, AdisaAzapagic "Life cycle environmental impacts of generating electricity and heat from biogas produced by anaerobic digestion", Energy volume 70 (2014) page from 181-193.

IOP Conf. Series: Materials Science and Engineering 1183 (2021) 012006 doi:10.1088/1757-899X/1183/1/012006

Modelling and Finite Element Analysis of an Aircraft Wing using Composite Laminates

Arijeet Nath[1], Raunika Anand[1], Jeet Desai[1], M.T.H. Sultan[2], S.Aravind Raj[1*]

1- Department of Manufacturing Engineering, School of Mechanical Engineering, Vellore Institute of Technology, Vellore, Tamil Nadu – 632014, India
2- Department of Aerospace Engineering, Faculty of Engineering, Universiti Putra Malaysia,43400 UPM Serdang, Selangor Darul Ehsan, Malaysia.

Corresponding author: aravindsakthivel@hotmail.com

Abstract. The use of finite element analysis is increasingly gaining popularity in the design and analysis of aircraft structures. Also, composite materials in the manufacturing and design of aircraft structures are increasing due to their exceptional properties of high stiffness to weight ratio. Applications include Composite structures are made up of laminates with different fibre orientations, which adversely affects the property of the composites formed. For this study, the three main parts of the aircraft wing are considered for designing: the spars, the ribs, and the aircraft's skin. The spars and made up of aluminium alloy, and the ribs are made up of structural steel. Two different wing models are designed. For Model 1, the skin is assigned Titanium alloy as a material, and for Model 2, CFRP (Carbon Fibre Reinforced Polymer) with resin epoxy is used as the skin material. The ribs are made of structural steel and the spars of titanium alloy for both models. This paper aims to compare the materials mentioned above to find the optimum strength to weight ratio. It was found that, for a given uniform load on the bottom of the aircraft's skin, a weight reduction of 2.37 % was observed in favour of Model 2, the deformation was reduced by 51 per cent, and the von-Mises stress was reduced by ≈ 85%. This paper's results can be used in manufacturing structural components of UAV, especially the wings due to lighter weight and more flexibility and durability than standard titanium alloys and the significant role played by the orientation of the plies in deciding the strength of the composite.

1. Introduction

Aircraft wings contribute a significant amount to the payload of any aircraft. It consists of many panels, and it is designed simultaneously to achieve an optimum design [1,2]. A good wing design involves the aerodynamic, structural design, as well as material selection [3]. The aerofoil wing shape is responsible for the generation of lift caused by the aircraft's forward velocity [4, 5, 6]. To increase the efficiency of an aircraft, it is essential to reduce the wing's weight by a significant amount, and it is a well-explored topic in the field of aircraft. Weight is a critical factor in aircraft design, and increasing stiffness, although it may reduce flutter onset, it does so at the expense of weight [7, 8, 9]. High strength and lightweight are two primary attributes of any aircraft wing [10, 11]. Materials used for aircraft components impact structural efficiency, payload energy consumption, cost, and reliability [1, 7, 12]. It has been observed that the most efficient way to increase the structural efficiency is by reducing the density, which in turn implies using lightweight materials [13,14]. Although certain alloys like Aluminium and Titanium help in a significant weight reduction of aircraft materials and

Content from this work may be used under the terms of the Creative Commons Attribution 3.0 licence. Any further distribution of this work must maintain attribution to the author(s) and the title of the work, journal citation and DOI.
Published under licence by IOP Publishing Ltd

have seen application in the aerospace industry for years, using composites helps with a massive weight reduction, thus facilitating the increased efficiency [7, 13, 15, 16]. A composite structure has laminates stacked together with different fibre orientation angles [13]. It consists of two or more constituent materials such as the elastic modulus or strength of each constituent is different [17, 18, 19, 20, 21]. The use of composites also provides better thermal stability and resistance to corrosion. The use of composite minimizes the structural weight, although it increases the manufacturing cost as a drawback [22]. There are various kinds of laminates; a unidirectional laminate is one with all the filaments arranged in one direction, a cross-ply laminate has layers of unidirectional laminates arranged at right angles with each other [23]. Composites like Carbon fibre are gaining popularity in aerospace, although, Aluminium is still leading the aircraft market [14, 24]. For example, A380 is a super-sized aircraft whose composition is 64% made of aluminium alloys. Similarly, for Boeing 777, another super-sized aircraft constitutes 70%. Aluminium in composition [7]. Analysis of laminate structures requires the comprehension of individual stresses and strains of the laminate, and therefore, the analysis of laminates requires the understanding of individual laminates [25, 26]. Kakur N et al. study the ILSS and flexural properties of CFRP and GFRP for five different laminate orientations and conclude through both experimental and theoretical investigation that modulus and flexural strengths are more significant 0° laminates. In contrast, the flexural strain is greater for (45°/-45°/45°) laminate layup [27]. Ernie I Basri et al. talk about composite ply orientation in the NACA 4415 wing, the structural analysis studies the layer-wise FEM for a Carbon Fiber Reinforced Polymer laminated composite in the different orientation of ply combination. The paper discusses how a specific ply orientation increases an aircraft wing [28]. Muhammad Amir Mirza et al. performed structural analysis over a three-dimensional wing of NACA 23015, which is identical to A320. The paper concludes that a wing with ribs at the root and the tip of the wing shows lesser deformation as compared to a wing that contains spar and wings [2].

For the current research, we have decided to optimize the weight of the wings with a NACA 23015 wing profile due to its similarities with A320 by changing the material composition and ply orientation. Airbus A320 is one of the most prominent and successful aeroplanes from the jetliner's family. Constructed after A300 and A310, A320 is the most commercially popular narrow-body airliner [29]. This aircraft was chosen because it is one of the most used commercial planes. This study compares the uses of Carbon fibre (395 GPa) with resin epoxy (CFRP) and Titanium alloy on the wing skin. A weight reduction was observed with the composite used compared to the alloy. Modelling of the aircraft wing was done using SolidWorks 2016. The composite layup is done on the FEA software ANSYS ACP PrePost module. Various layups of the composite material (CFRP) are done to find a configuration with the least total deformation and equivalent (Von-Mises) stress. The layups are done following the Classical Laminate Theory (CLT).

The following study employs a definite structure employing sections and subsections. The methodology is discussed in the next section, in which the geometrical configuration, material specifications, mesh generation, loading conditions, and finally, the layup of composite materials are discussed followed by the results and discussions in section 3 followed by the conclusion in section 4.

2. Methodology

2.1 Geometrical Configuration

The aircraft wing design is a complicated iterative procedure. Besides, the calculations are commonly repeated quite a lot of times. For this wing's design, two spar wing design is considered suitable as this configuration is used for most commercial aircraft construction. The NACA profile 23015 is used, referencing the Airbus A320. The model is primarily composed of two spars, ten ribs, and one part of the skin. The design was sketched and assembled in SolidWorks 2016, and the FEM process was done in ANSYS Workbench. The wing's span is set to 4.5 m, and the chord length of the wing is 1 m. The root and tip chord thickness is 0.04 m, and all other rib thickness is kept to 0.02 m. The design specifications are depicted in Figure 1 and Figure 2.

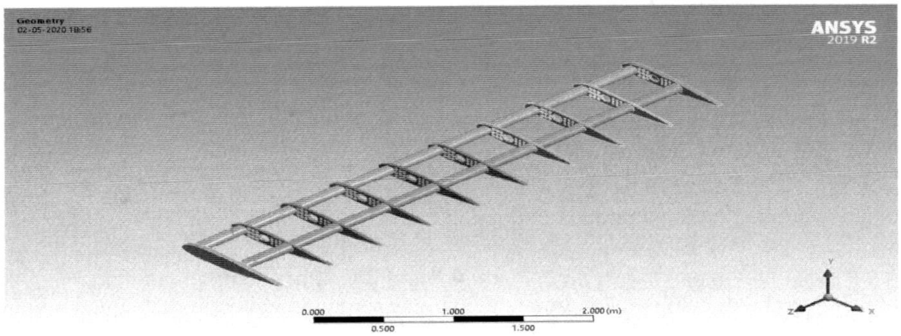

Figure 1. Internal structure of the wing

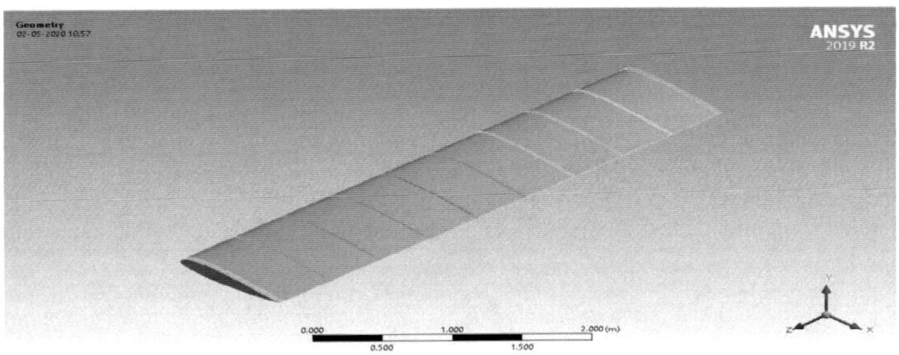

Figure 2. External Structure of the wing

2.2 Material Specifications

The specifications of the different parts of the model are mentioned below. The construction of composite material is done using Carbon Fibre (395 GPa) and Resin Epoxy as the global drop-off material. The details of the properties of the materials used are given in the tables below:

Table 1. Titanium Alloy

Properties	Values	Unit
Density	4620	Kg/m3
Young's Modulus	9.6E+10	Pa
Poisson's ratio	0.36	
Tensile Yield Strength	9.3E+08	Pa
Tensile Ultimate Strength	1.07E+09	Pa

Table 2. Structural Steel

Properties	Values	Unit
Density	7850	Kg/m3
Young's Modulus	2E+11	Pa
Poisson's ratio	0.3	
Tensile Yield Strength	2.5E+08	Pa
Tensile Ultimate Strength	4.6E+09	Pa

Table 3. Carbon Fibre 395 GPa (fibre only)

Properties	Values	Unit
Density	1800	Kg/m3
Young's Modulus X direction	3.95E+11	Pa
Young's Modulus Y direction	6E+09	Pa
Young's Modulus Z direction	6E+09	Pa
Poisson's Ratio XY	0.2	
Poisson's Ratio YZ	0.4	
Poisson's Ratio XZ	0.2	
Shear Modulus XY	8E+09	Pa
Shear Modulus YZ	2.1429E+09	Pa
Shear Modulus XZ	8E+09	Pa

Table 4. Mechanical Properties of Model 1 and 2, Elements and Materials

Sr. No	Model	Ribs	Spars	Skin	Weight(in kgs)
1	Model 1	Structural Steel	Titanium Alloy	Titanium Alloy	540.38
2	Model 2	Structural Steel	Titanium Alloy	Carbon Fibre(395 GPa) Epoxy Composite	527.56

2.3 Mesh generation

Mesh generation is of prime importance in FE analysis considering the results generated are dependent on the quality of the mesh generated. Derived quantities such as stresses and strains are calculated at the element integration points and extrapolated to the nodes. The value described in ANSYS Workbench is the average value at the node, which is based on all elements connected to that node, and the difference in averaged and un-averaged results is a mesh discretization error indication. As mentioned by Kanesan et al., the aspect ratio of the mesh generated should also be less. Since the loading will affect the skin directly, meshing is much more refined for the skin as compared to other parts of the model to save computational time. Also, the mesh should accurately represent the geometry and computational loads and domain. [29].Some important steps were undertaken to ensure a better mesh generation for the study:

1. Face meshing was applied to the top and bottom half of the skin, and mesh defeaturing was set to yes for the whole mesh to remove unwanted parts in the model, thereby reducing calculation time.
2. Body Sizing was applied to the spars and the ribs
3. Capture proximity and capture curvature were set to Yes

The total number of nodes and elements generated was 131978 and 54019.

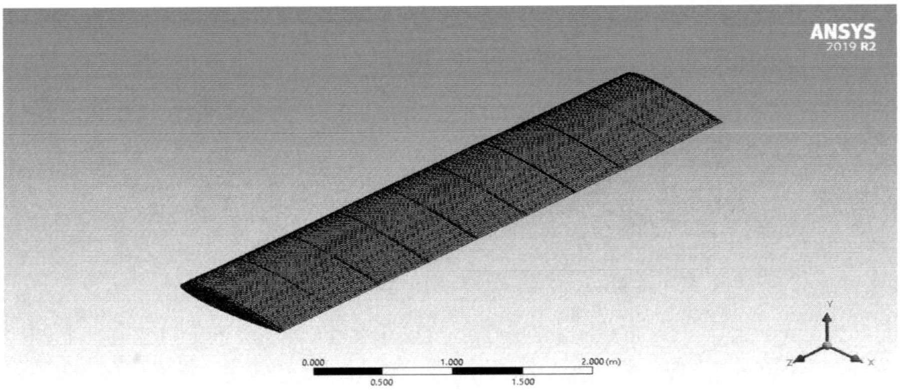

Figure 3. Mesh Generation in the outer skin

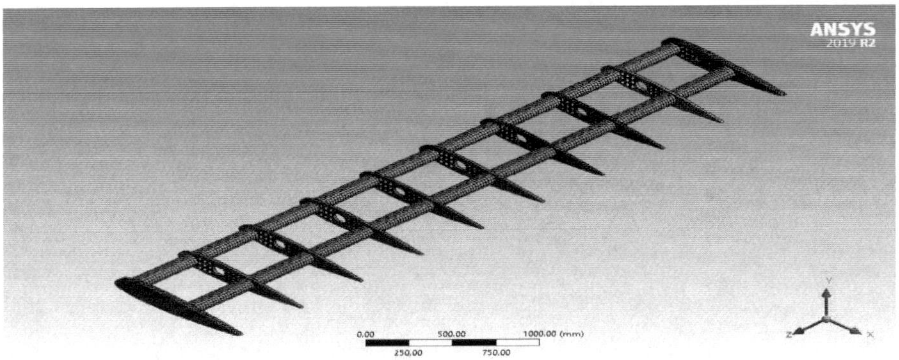

Figure 4. Mesh Generation in the inner ribs and spars

2.4 Boundary and Loading Conditions

2.4.1 Boundary Conditions: As far as the loading conditions are concerned, the fuselage was not included in the model. The boundary conditions consider that an aircraft's wing model is connected to the fuselage, which is not included in the study. The wing acts as a cantilever beam connected to the fuselage [23]. Hence the face attached to the fuselage is constrained in all the DOFs i.e. UX = UY = UZ = ROTX = ROTY = ROTZ =0.

2.4.2 Loading Conditions: The loading condition for this study is assumed to be of a wing torque box under the influence of aerodynamic loading [30]. The loads' calculation is done by assuming a uniformly distributed pressure load applied in the vertical direction on the bottom part of the wing. A vertical pressure of 1963.770304 Pa is applied to the bottom part of the skin.

The aerodynamic loading conditions can be calculated using the formula given in eq. (1).

$$L \div (0.4 \times W) = FOS \times \eta_{max} \qquad (1)$$

$L \div (0.4 \times W) = FOS \times \eta_{max}$ (L=lift capacity, W =weight of the aircraft.)

Referencing from [2] FOS is taken to be 1.5

$$W = 540.38 \times 9.81 = 5301.1278N \qquad (2)$$

In general, 80% of the total lift load is assumed to be uniformly distributed on the bottom part of the wing with the rest of the 20% being attached to the fuselage. This factors in the value of 0.8 in eq. (3).

$$\therefore \text{Lwing} = 3.5\times1.5\times0.4\times5301.1278\times0.8=8905.894704\text{N} \quad (3)$$

$$\therefore \text{P} = (\text{Lwing}\div\text{S}) = (8905.894704\div4.5351) = 1963.770304\text{Pa} \quad (4)$$

(S = Surface Area of the wing part where the load is applied)

The loading and boundary conditions are shown in the fig. 5 below:

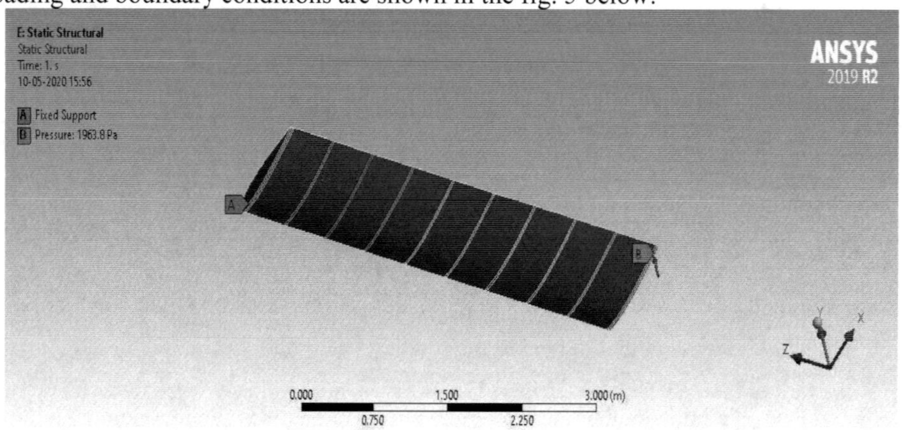

Figure 5. Boundary and Loading Conditions on the wing

2.5 Modelling of composite layers for the Skin

The composite layup is done on the FEA software ANSYS ACP Pre module. Various layups of the composite material (CFRP) are done to find a configuration with least total deformation, equivalent stress. The layups are done by the classical laminate theory (CLT). Using CLT, we can find the structural response of a composite laminate given an arbitrary set of loads and a stacking sequence [31]

Classical Lamination Theory predicts the behavior of the laminate within the framework of the following assumptions [32]:

1. Each laminate layer is assumed to be orthotropic as well as quasi-homogeneous.
2. The laminate is thin with the lateral dimensions significantly larger than the thickness and that the loads applied on the laminate is only in in-plane directions.
3. Out of plane displacements are assumed to be insignificant compared to the laminate thickness, and in-plane displacements vary linearly throughout the laminate thickness.
4. All displacements are insignificant compared to the laminate thickness, and perpendicular distances from the middle surface stay constant.
5. The relations between applied forces and moments at an arbitrary point (x, y), and the resultant mid-plane curvatures and strains, can be summarized as follows in fig. 6. The A/B/B/D matrix in brackets is the laminate stiffness matrix, and its inverse is the laminate compliance matrix.

Relation between applied forces and moments, and the resultant mid-plane curvatures and strains

$$\begin{Bmatrix} N \\ M \end{Bmatrix} = \begin{bmatrix} \mathcal{A} & \mathcal{B} \\ \mathcal{B} & \mathcal{D} \end{bmatrix} \begin{Bmatrix} \epsilon^0 \\ \kappa \end{Bmatrix}$$

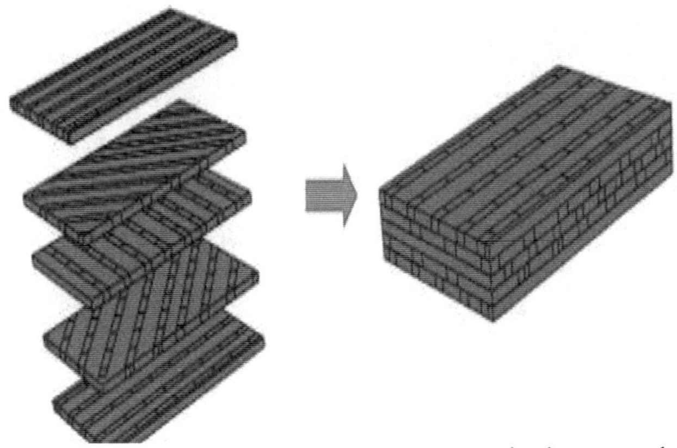

Figure 6. Ply arrangement of various arrangements in a composite layup made in Solidworks [33]

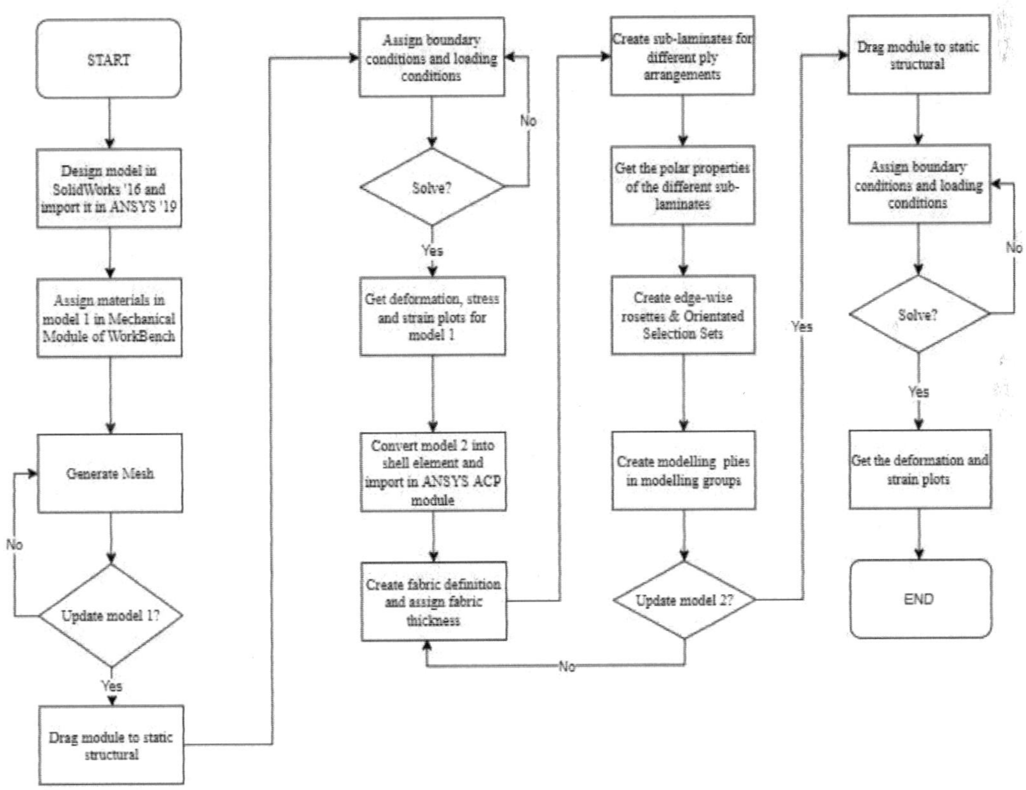

Figure 7. Flowchart representing design and analysis process of the study

3. Methodology

3.1. Validation of the paper with previously published results

The model in this paper is validated with Zakuan, M.A.M.B.M [2]. For this validation, aluminium alloy is used in the ribs and wing skin, and titanium alloy is used for primary and secondary spars.

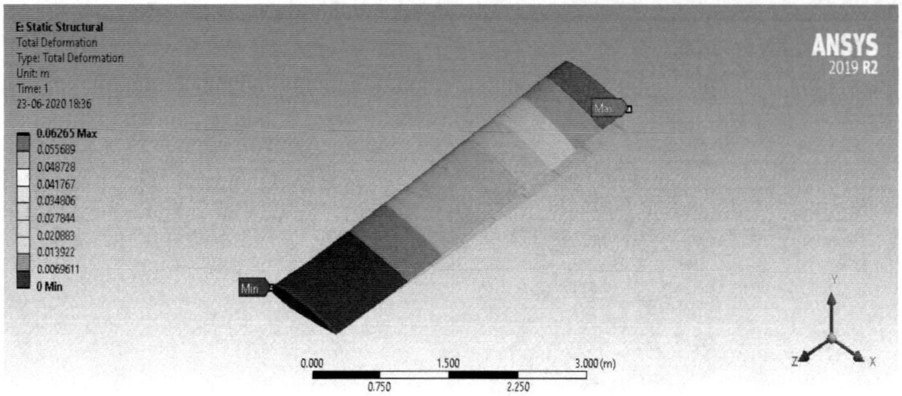

Figure 8. Total Deformation

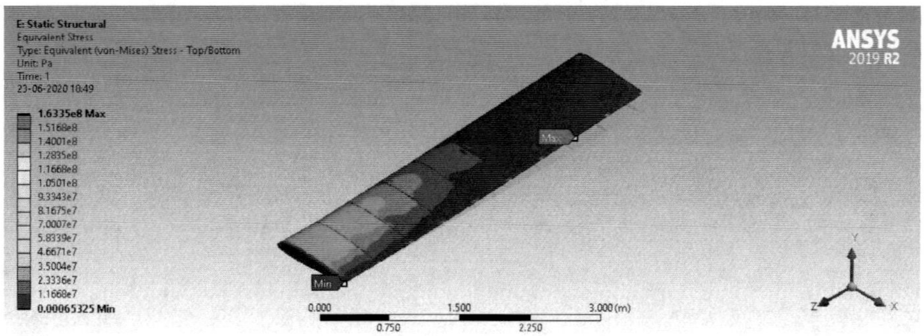

Figure 9. Equivalent (Von-Mises) Stress

Table 5. Validation with published data

S.No.	Total Deformation(m)	Equivalent(von-Mises) Stress(Pa)
Obtained Results	0.06265	1.6335e8
Existing Data	0.10126	1.9387e8

As seen from the table, the results in total deformation, equivalent stress between the obtained results and the existing data are relatively close to each other, and hence the said model can be used for further optimization process using composites. The slight differences in the results are due to differences in meshing parameters and the design of the wing used for this study being optimized to reduce deformation values. Hence as can be inferred from the table, this design can be further used to optimize the wing model using CFRP with different ply orientations.

3.2. Static Structural Analysis of Model 1 and Model 2

Static Structural analysis of Model 1 and 2 is done to find the total deformation and the Von-Mises stress plots.

3.2.1. *Analysis of Model 1:* Model 1 has titanium alloy as the added material for the skin of the wing as well as the spars and structural steel as the material for the ribs.

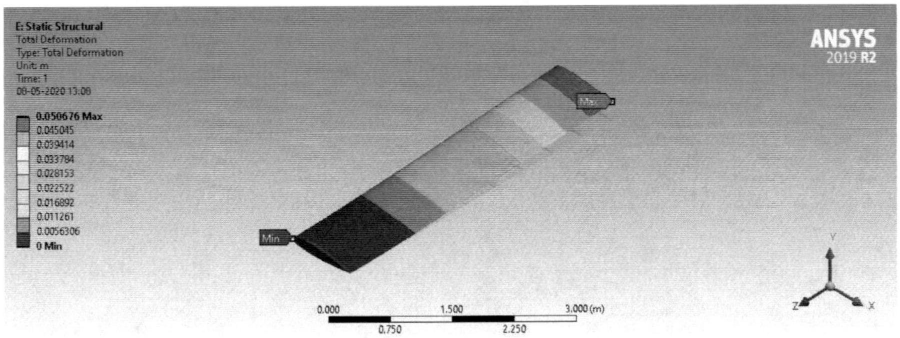

Figure 10. Total deformation of Model 1

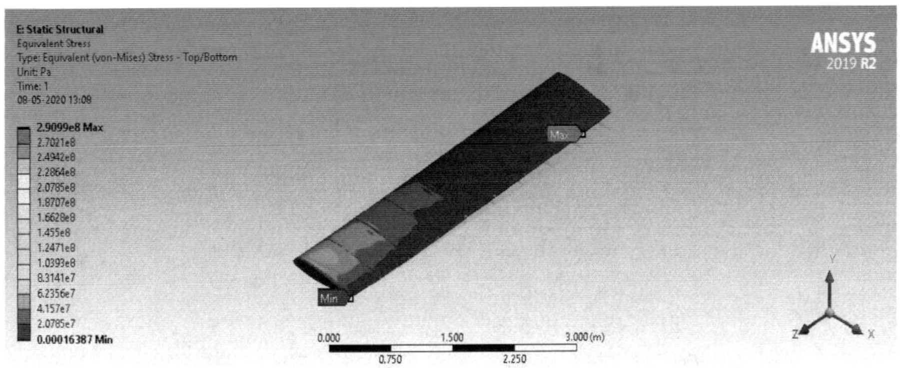

Figure 11. Equivalent (Von-Mises) Stress of Model 1

As seen in the figure, the wing acts as a cantilever beam with the root attached to the fuselage. The spar primarily carries the model's load. They are the primary members that provide structural support for the wing against twisting (torsion) and upward bending forces while the wing generates lift. On the other hand, Ribs, the internal structural parts running fore and aft in the wing, provide the airfoil shape to the structure. The maximum deformation, in this case, occurs with the value of 0.050676 m at the tip of the wing structure, while the minimum deformation is at the root part of the wing structure.

Most of the stress concentration occurs at the ribs' tip, which means these parts are most prone to damage and have a reduced life compared to the other components. Also, the stress values gradually increase as we move away from the part attached to the fuselage. This is to be expected as the wing behaves similar to a cantilever beam. The maximum stress value is 2.9099e+8 Pa.

3.2.2. *Analysis of Model 2:* Model 2 employs the use of composite layups with the intent to reduce deformation, stress values, and reducing the weight of the wing. A laminate code is generally used to identify the lamina stacking sequence in a laminate. Knowing the code assists with comparing with the form of [A], [B], and [D] before analysis [8]. This study has employed different types of lamina structures such as symmetric, anti-symmetric, and quasi-isotropic laminates to find which laminate orientation is best suited in terms of strength, stiffness, and bending matrices

Model 2.A: A ply sequence of [0/90/0/90/0]s is applied on the skin of the wing made up of Carbon-fibre (395 GPa)-epoxy material. In such a symmetric layup, the [Bi,j] matrix becomes zero which means bending and extension are uncoupled, allowing simplified solution procedures.

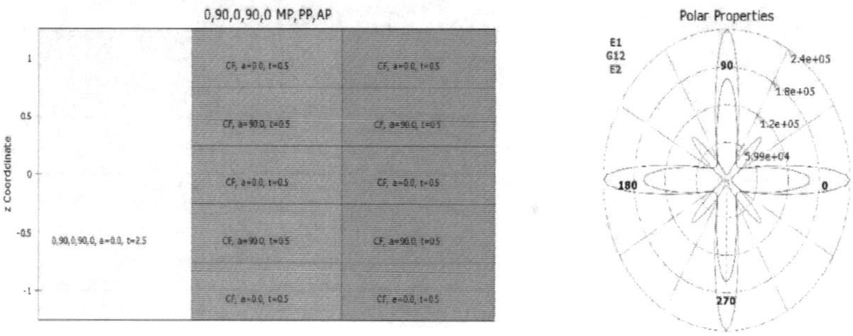

Figure 12. Polar Coordinates of [0/90/0/90/0]

Table 6. Laminate Engineering Constants of [0/90/0/90/0]

Constant	Value
Flexural Laminate Shear Stiffness	8000
Flexural Laminate Stiffness E1	314262.4
Flexural Laminate Stiffness E2	86960.25
Laminate Shear Stiffness	8000
Laminate Stiffness E1	239536.6
Laminate Stiffness E2	161692.2
Out of Plane Shear	1715.785
Out of Plane Shear	2923.015
Shear Correction Factor k44	0.3825
Shear Correction Factor k55	0.516695

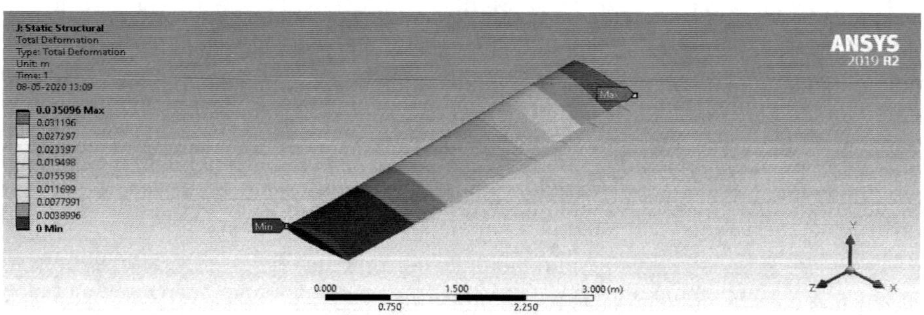

Figure 13. Total Deformation of [0/90/0/90/0]

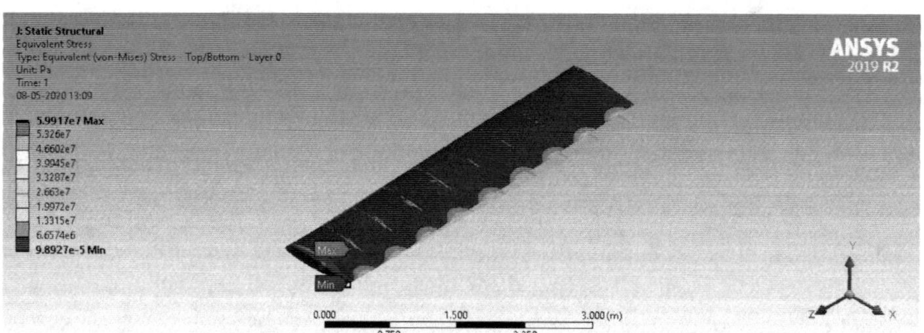

Figure 14. Von-Mises stress of [0/90/0/90/0]

The total deformation plot is similar to what was seen in Model 1. The maximum deformation value is 0.035096 m which occurs at the root of the wing. Here, the maximum stress value is 5.9917e+7 Pa and acts on the skin of the model. The minimum value is 9.8927e-5 Pa at the root of the wing.

Model 2.B: A ply sequence of [0/-35/45/90/-45/0/45/90/-45/35/0] is applied on the skin. This layup is a general type of laminate which is antisymmetric about the middle ply and not the midplane. Each ϴ is associated with a -ϴThe [Bi,j] matrix is not zero.

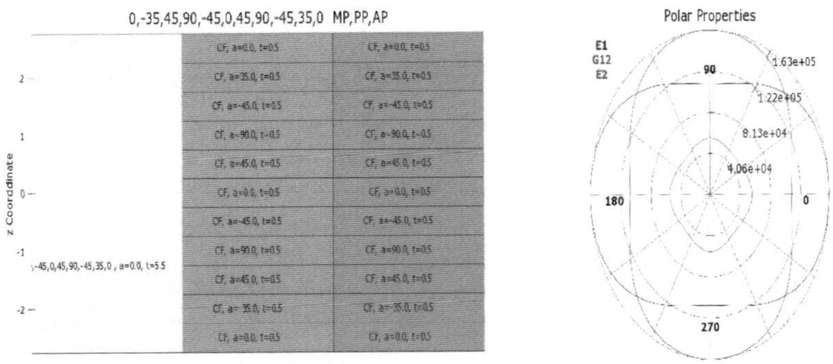

Figure 15. Polar Co-ordinates of [0/-35/45/90/-45/0/45/90/-45/35/0]

Table 7. Laminate Engineering Constants of [0/-35/45/90/-45/0/45/90/-45/35/0]

Constant	Value
Flexural Laminate Shear Stiffness	48297.99
Flexural Laminate Stiffness E1	227833.6
Flexural Laminate Stiffness E2	60384.85
Laminate Shear Stiffness	56073.39
Laminate Stiffness E1	162564.5
Laminate Stiffness E2	109438.3
Out of Plane Shear	1890.714
Out of Plane Shear	2840.326
Shear Correction Factor k44	0.408973
Shear Correction Factor k55	0.514573

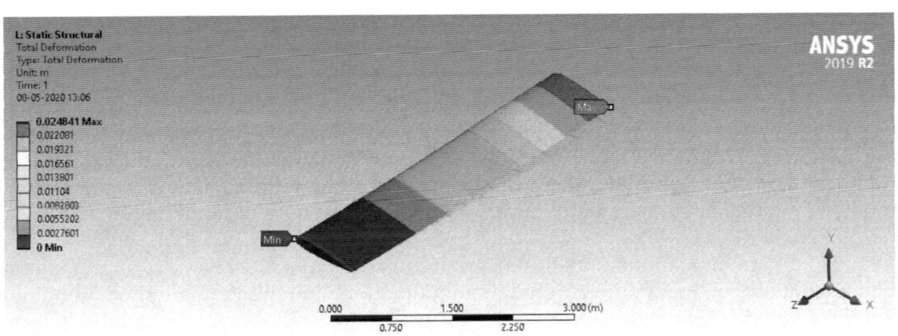

Figure 16. Total Deformation of [0/-35/45/90/-45/0/45/90/-45/35/0]

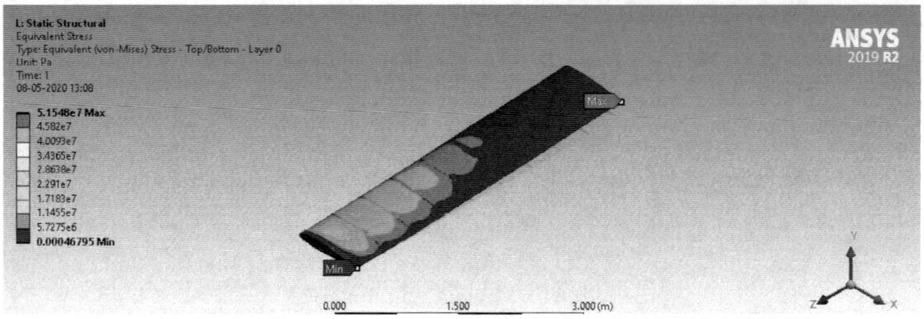

Figure 17. Von-Mises Stress of [0/-35/45/90/-45/0/45/90/-45/35/0]

The maximum total deformation of the model in this arrangement of plies is 0.024841 m. Looking at the stress plot, we see that the maximum stress value is 5.1548e+7 Pa while the minimum value is 0.00046795 Pa on the root of the wing.

Model 2. C: This model has a ply sequence of [0/90/45/-45/0/-45/45/90/0] which is a quasi-isotropic laminate layup. A laminate is known as a quasi-isotropic laminate when its extensional-stiffness matrix [Ai, j] behaves like an isotropic material. The reason it is called "quasi-isotropic" is that [Bi, j] and[Di, j] may or may not behave like an isotropic material. For this ply sequence, the stiffness is independent of the rotation angle of the laminate and that the extension and shear matrices are uncoupled.

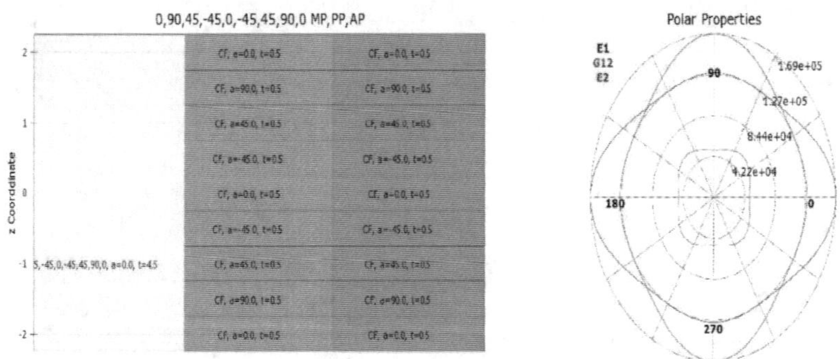

Figure 18. Polar Co-ordinates of [0/90/45/-45/0/-45/45/90/0]

Table 8. Laminate Engineering Constants

Constant	Value
Flexural Laminate Shear Stiffness	22625.9
Flexural Laminate Stiffness E1	225004.5
Flexural Laminate Stiffness E2	135257.1
Laminate Shear Stiffness	48760.26
Laminate Stiffness E1	168748.9
Laminate Stiffness E2	128535.7
Out of Plane Shear	2070.831
Out of Plane Shear	3008.249
Shear Correction Factor k44	0.436329
Shear Correction Factor k55	0.557411

IOP Conf. Series: Materials Science and Engineering 1183 (2021) 012006 doi:10.1088/1757-899X/1183/1/012006

Figure 19. Total Deformation for [0/90/45/-45/0/-45/45/90/0]

Figure 20. Von-Mises Stress for [0/90/45/-45/0/-45/45/90/0]

As seen in the deformation plot, the maximum deformation value is 0.025436m. The maximum value of stress is found to be 4.2801e+7 Pa.

Table 9. Results of the analysis on the wing model

Model Types	Skin Material	Ply orientation		Total Deformation (m)	Von-Mises Stress (Pa)
Model 1	Titanium Alloy	Nil		0.050676	2.9099e+8
Model 2	CFRP	A	[0/90/0/90/0]	0.035096	5.9917e+7
	CFRP	B	[0/-35/45/90/-45/0/45/90/-45/35/0]	0.024841	5.1548e+7
	CFRP	C	[0/90/45/-45/0/-45/45/90/0]	0.025436	4.2801e+7

4. Conclusion

A comparison was drawn between two models of the aircraft wing with similar loading and boundary condition and similar design with change in materials of the skin. Titanium alloy in Model 1 and Carbon-Fibre Epoxy Composite material (CFRP) in Model 2 and the effect of different ply orientations were evaluated. Reading the results of the different ply orientations in Model 2, the lowest deformation of 0.024841 m is observed in Model 2.B which has the stacking sequence of [0/-35/45/90/-45/0/45/90/-45/35/0]. On comparing the deformation in Model 1 and Model 2.B, a weight reduction of 2.37 % is observed in the wing favouring Model 2.B, and the total deformation is reduced by $\approx$ 51%. Similarly, the von-Mises stress is reduced by $\approx$ 85%. This underscores the fact that the application of composite materials on the skin of the wing compared to standard titanium alloys provide a higher stiffness to weight ratio. The orientations of plies are of prime importance in deciding the strength of the composite stack up. Various stack-up orientations are used through iterative procedures and compared. It is found that model 2.B is the optimum choice for composite layups among the laminate orientations in the study because it showcased better strength to weight ratio than others because of lesser deformation and stress value. Further study can be focused on making cost-effective materials and optimising the design of the wing to achieve optimum strength to weight ratio. Better and more accurate results can be obtained by generating a finer mesh, and composite failure analysis can be done as an extension to this study.

5. Nomenclature

N	Traction at point (x, y)
M	Moment at point (x, y)
[Ai,j]	Extensional-stiffness matrix
[Bi,j]	Extension-bending coupling stiffness matrix
[Di,j]	Bending-stiffness matrix
ϵ^0	The mid-plane strain and κ is the vector comprising of second derivatives of the displacement, known as the curvature.

6. References

[1] Dababneh, O. and Kipouros, T., 2018. A review of aircraft wing mass estimation methods. Aerospace Science and Technology, 72, pp.256-266.

[2]Zakuan, M.A.M.B.M., Aabid, A. and Khan, S.A., Modelling and Structural Analysis of Three-Dimensional Wing.

[3] Boutemedjet, A., Samardžić, M., Rebhi, L., Rajić, Z. and Mouada, T., 2019. UAV aerodynamic design involving genetic algorithm and artificial neural network for wing preliminary computation. Aerospace Science and Technology, 84, pp.464-483.

[4]Venkatesan, S.P., Beemkumar, N., Jayaprabakar, J. and Kadiresh, P.N., 2018. Modelling and analysis of aircraft wing with and without winglet. International Journal of Ambient Energy, pp.1-11.

[5] Kulshreshtha, A., Gupta, S.K. and Singhal, P., 2020. FEM/CFD analysis of wings at different angle of attack. Materials Today: Proceedings.

[6] Verma, G., Sharma, A. and Prajapati, Y.K., 2020. A Numerical Study to Investigate the Modal Analyses of Cracked Airplane Wing (NACA2415). In Advances in Applied Mechanical Engineering (pp. 405-411). Springer, Singapore.

[7] Zhu, L., Li, N. and Childs, P.R.N., 2018. Light-weighting in aerospace component and system design. Propulsion and Power Research, 7(2), pp.103-119.

[8] Taub, A.I. and Luo, A.A., 2015. Advanced lightweight materials and manufacturing processes for automotive applications. Mrs Bulletin, 40(12), pp.1045-1054.

[9] Hussin, M.S., Ghorab, A. and El-Samanoudy, M.A., 2018. Computational analysis of two-dimensional wing aeroelastic flutter using Navier-Stokes model. Ain Shams Engineering Journal, 9(4), pp.3459-3472.

[10] Das, S.K. and Roy, S., 2018, August. Finite element analysis of aircraft wing using carbon fiber reinforced polymer and glass fiber reinforced polymer. In IOP Conf. Ser. Mater. Sci. Eng (Vol. 402, p. 012077).

[11] Miracle, D., 2019. Lightweighting and the Future of Aerospace Metals. In Light Weighting for Defense, Aerospace, and Transportation (pp. 27-38). Springer, Singapore.

[12] Nair, C.K., 2019. Innovations in Materials and Engineering for Light-Weighting Aircraft. In Light Weighting for Defense, Aerospace, and Transportation (pp. 73-88). Springer, Singapore.

[13] Shabeer, K.P. and Murtaza, M.A., 2013. Optimization of aircraft wing with composite material. International Journal of Innovative Research in Science, Engineering and Technology, 2(6).

[14] Achunala Rajesh, Shivappa H A, Performance Evaluation of Aircraft Wing for Different Lamina Orientation of Laminated Composite, International Journal for Ignited Minds (IJIMIINDS), (2016)

[15] Liu, D., Tang, Y. and Cong, W.L., 2012. A review of mechanical drilling for composite laminates. Composite structures, 94(4), pp.1265-1279.
[16] Ghazali, E., Dano, M.L., Gakwaya, A. and Amyot, C.O., 2019. Mechanical performance of repaired sandwich panels: Experimental characterization and finite-element modelling. Journal of Sandwich Structures & Materials, 21(4), pp.1357-1378.
[17] Staab, G., 2015. Laminar composites. Butterworth-Heinemann.
[18]Alieldin, S.S., Alshorbagy, A.E. and Shaat, M., 2011. A first-order shear deformation finite element model for elastostatic analysis of laminated composite plates and the equivalent functionally graded plates. Ain Shams Engineering Journal, 2(1), pp.53-62.
[19] Pramanik, A., Basak, A.K., Dong, Y., Sarker, P.K., Uddin, M.S., Littlefair, G., Dixit, A.R. and Chattopadhyaya, S., 2017. Joining of carbon fibre reinforced polymer (CFRP) composites and aluminium alloys–A review. Composites Part A: Applied Science and Manufacturing, 101, pp.1-29.
[20] Naveen, J., Jawaid, M., Vasanthanathan, A. and Chandrasekar, M., 2019. Finite element analysis of natural fiber-reinforced polymer composites. In Modelling of Damage Processes in Biocomposites, Fibre-Reinforced Composites and Hybrid Composites (pp. 153-170). Woodhead Publishing.
[21] Ozkan, D., Gok, M.S. and Karaoglanli, A.C., 2020. Carbon Fiber Reinforced Polymer (CFRP) Composite Materials, Their Characteristic Properties, Industrial Application Areas and Their Machinability. In Engineering Design Applications III (pp. 235-253). Springer, Cham.
[22] Kaufmann, M., Zenkert, D. and Mattei, C., 2008. Cost optimization of composite aircraft structures including variable laminate qualities. Composites Science and Technology, 68(13), pp.2748-2754.
[23] Rajadurai, M., Vinayagam, P., Priya, G.M. and Balakrishnan, K., 2017. Optimization of ply orientation of different composite materials for aircraft wing. International Journal of Advanced Engineering Research and Science, 4(6), p.237200.
[24] Wang, G.D. and Melly, S.K., 2018. Three-dimensional finite element modeling of drilling CFRP composites using Abaqus/CAE: a review. The International Journal of Advanced Manufacturing Technology, 94(1-4), pp.599-614.
[25] Koruche, U.S. and Patil, S.F., 2015. Application of classical lamination theory and analytical modeling of laminates. International Research Journal of Engineering and Technology, 2(02), p.958.
[26] YazdaniSarvestani, H. and Naghashpour, A., 2014. Analysis of free edge stresses in composite laminates using higher order theories. Indian Journal of Materials Science, 2014.
[27] Naresh, K., Krishnapillai, S. and Velmurugan, R., 2017. Effect of fiber orientation on carbon/epoxy and glass/epoxy composites subjected to shear and bending. In Solid State Phenomena (Vol. 267, pp. 103-108). Trans Tech Publications Ltd.
[28] Basri, E.I., Sultan, M.T., Faizal, M., Basri, A.A., Abas, M.F., Majid, M.A., Mandeep, J.S. and Ahmad, K.A., 2019. Performance analysis of composite ply orientation in aeronautical application of unmanned aerial vehicle (UAV) NACA4415 wing. Journal of Materials Research and Technology, 8(5), pp.3822-3834.
[29] Kanesan, G., Mansor, S. and Abdul-Latif, A., 2014. Validation of UAV wing structural model for finite element analysis. JurnalTeknologi, 71(2).
[30] Ünlüsoy, L., 2010. Structural design and analysis of the mission adaptive wings of an unmanned aerial vehicle (Master's thesis).
[31] Giurgiutiu, V., 2015. Structural health monitoring of aerospace composites. Academic Press.
[32] Stuart Dutton, Donald Kelly, Alan Baker, Composite Materials for Aircraft Structures Second Edition (2012).
[33]Strength Criteria of Composite Material supported by FEM analysis, 17th February, 2019. <http://www.aerospacengineering.net/strength-criteria-of-composite-material-supported-by-fem-analysis/>

Polyelectrolyte Complex for Drug Delivery in Biomedical Applications: A Review

Pratik Jagtap[1*], Kiran Patil[2], Pankaj Dhatrak[3]

[1]UG Student, School of Chemical Engineering, Dr. Vishwanathkarad, MIT-World Peace University, Pune-38, India.
[2]Faculty of Chemical Engineering, Dr. Vishwanathkarad, MIT-World Peace University, Pune-38, India.
[3]Faculty of Mechanical Engineering, Dr. Vishwanathkarad, MIT-World Peace University, Pune-38, India.

[*]Corresponding author Email: pratikjagtap992@gmail.com

Abstract. There are so many significant developments made in drug delivery system (DDS) in which Polymer DDS has reached great developments. The main motive of this paper is to focus on DDS developing by Polyelectrolytes. Polymers have achieved much importance in DDS. Polyelectrolytes shows an exceptional category of polymer compounds containing of opposite charged polyions which are mainly cationic or anionic charges; they are hydrophilic in nature because of presence of opposite charge, can be used in various of applications in DDS. Polyelectrolytes (PE) exhibits various types of reactive groups that permits easy, adaptable modification with functional agents such as targeting molecules. Further, the oppositely charged nature exist in polyelectrolytes been used to transport macromolecules such as DNA and nanoparticles. Significance of smart polyelectrolyte is growing day by day from last several year. In the last several years, polyelectrolytes representing promising means for targeted based drug delivery and gene treatment. In the present paper, polyelectrolyte, polyelectrolyte complexes, deposition method for multilayer system, features, characterization of polyelectrolytes, and applications are reviewed.

Keywords: *Chitosan, DNA, Drug Delivery, Polyelectrolyte, Macromolecular characterization*

1. Introduction

Rapid Researchers have focused on designing smart drug carriers to enhance medical treatment using "on demand" targeting approaches. Using such an approach, a smart drug vehicle carriers therapeutic agents to be released at targeted sites with minimal damage to healthy cells [1]. Polymeric material is extremely attractive to specific conditions. According to pH, temperature, magnetic and electric fields have become extremely attractive in DDS. In DDS, factors such as toxicity, biocompatibility, degradability, and sustainability of releases are considered and therefore the design of systems is usually so much more complicated. Polyelectrolyte have emerged as potential materials for creating these carriers due to their tunable size and biocompatibility. Polyelectrolytes hold great features, such

Content from this work may be used under the terms of the Creative Commons Attribution 3.0 licence. Any further distribution of this work must maintain attribution to the author(s) and the title of the work, journal citation and DOI.
Published under licence by IOP Publishing Ltd

as fine-tuning of thickness, stiffness, stability, morphology, and topography; they may exhibit special biological properties, such as muco adhesion. Weak PE is permissible for a varied range and they are useful in the formation of "smart" materials for chemical separations [2]. For the selection of polymer material, it is important to select the material, which can control the encapsulation efficiency, rate of deliver of drug, and time required to release the drug. colloidal transporters (micro, nanoparticles, micelles), implantable networks, and polymeric medicine these are the three main classes observed in the polymeric DDS[3]. Polyelectrolytes and functionalized polymers presently play an essential part in the development of DDS by providing precise transport of remedial agent.

Over the last 20 years, statistics as a discipline has expanded tremendously. Statistics has a role whenever there is a demand for quantification. Statisticians are a key collaborator in the creation of new pharmaceuticals because of their capacity to conceive in terms of variability, to extract signals from noise, to control sources of bias and variation, and to optimise under given conditions. For conventional, sustained release, and controlled release DDSs, changes in plasma drug concentrations over time [4].

2. Drug delivery system (DDS)

Drug delivery is a means of administering medicine in such a way that the medication concentration in some sections of the body is higher than in others. Drug delivery aims to concentrate drug in the tissues of interest while lowering its relative concentration in the remainder of the body. This increases the efficacy of the while lowering the adverse effects. The delivery of drugs to receptors, organs, or any other specified portion of the body to which the drugs to be delivered is known as drug targeting exclusively[5]. The therapeutic index of a drug is a measure of how effective it is at treating a specific condition as determined by its pharmacological response Access and specific introduction are essential for safety of the drug and its potential receptor. The various parameter used in the DDS is explained below.

2.1 Controlled release

Maintaining the concentration within the therapeutic window can improve drug efficacy (effective dose). By managing drug diffusion, the rate of dissolution, or carrier degradation, polymer carriers loaded with medicines provide controlled temporal and spatial release of a drug[6].

2.2 Targeted delivery

Localization at the organ, tissue, cellular, or organelle level can improve drug efficacy while reducing toxicity. Coating or conjugating the carrier with affinity reagents allows for targeting nucleic acids, peptides, antibodies, and other molecules that bind to specific targets. Proteins, nucleic acids, and polysaccharides are all examples of cell receptors[7].

2.3 Solubility enhancement

Low drug solubility and stability can significantly limit the efficacy of a potential treatment candidate. To increase the in vivo solubility of lipophilic and hydrophobic drugs, drug delivery systems can be created. Hydrophobic medicines can be delivered either via encapsulating in a drug delivery carrier or by injection[8].

3. Polyelectrolyte

Most of the highly charged molecules Polyelectrolyte when placed in ionized solvent, a macromolecular material that has repeating units and dissociates into highly charged polymeric molecule. Polyelectrolytes is a polymer of electrolyte with several repeating chain units or bear repeating units of electrolyte group [9].The study of the polyelectrolyte has considerable potential in the field of controlled and targeted delivery. Polyelectrolytes extensively used in controlled DDS to stabilise drugs and alter the drug release features. The opposite charge present on the repeating units of the polyelectrolyte is nullified by oppositely charged ions which can be able to form electro neautrality[10]. Polyelectrolytes has particular advantages over the non-polar polymers for example 1) In aqueous solution, the method of coating can be accompanied and it helps to reserve the bioactivity of fragile proteins. 2) By use of the polyelectrolyte, it allows the control of the release behaviour by choosing suitable polymers. polyelectrolytes are capable materials for the development of the controlled DDS[11].

3.1 Polyelectrolyte Complex (PECs)
One of the significantfeaturesof polyelectrolyte is that capability of the formation of the complex, when mixed with other the opposite charged polyelectrolyte. (PECs) and polyelectrolyte multilayers contributes similar physical, chemical type of properties in form of their internal, physical structure also morphology[12].PECs have the capacity of the formation of the complexes with multiple various oppositely charged particles. Interactions between polycations and polyanions (Due to electrostatic force of attraction) over the mixing of aqueous solutions of the opposite charged PEs, it tends to produce dense phase which is different from the solvent[13]. Considering factors such as origin, composition and molecular construction, polyelectrolytes are classified. Some of main PE are: 1) natural PE (Nucleic acid, pectin, Carrageenan, Alginate(refined from brown seaweed)). 2) Chemically adapted PE (Pectin, Chitin), 3) artificial polymer (Polyvinyl ,poly benzene trialkyl ammonium, Poly vinyl sulfonic acid, Poly acrylic or methacrylic acid), Poly styrene (sulfonic acid, Poly acryl amido alkyl trialkyl ammonium)[14].

4.Characterization of PECs
Depending on macromolecular and electrochemical properties, polyelectrolyte properties differentiated in two direction.

4.1 Macromolecular characterization
4.1.1. Chromatographic method. Field-flow fractionation, electrophoresis and size exclusion are some macromolecular properties of PECs. Main beliefs of selective polyelectrolyte coacervation might be functional to chromatography, if the polyelectrolyte were restrained on a chromatographic support part[15].

4.1.2. Osmatic Pressure measurement. Molecular mass of PEs can identify using osmatic pressure measurements (OPM).Because of their dissociation nature, osmotic pressure of the Polyelectrolytes differs frequently. Osmotic pressure can be find by a pressure difference between the sample section and a section with pure solvent or an electrolyte solution (OPM)[16]. At low salt concentration polyelectrolyte solution displays its exclusive feature as its osmotic pressure considerably go beyond the osmotic pressure of discharged polymers at the equal polymer concentrations[17].

4.1.3. Light scattering method. To illustrate Polyelectrolytes in solution form, light scattering systems are commonly used. This technique can deliver complete data on particle size of PE, shape of PE and mass of the scattering elements, particle solvent contact and polydispersity[18].polyelectrolyte solution properties have been studied mainly by viscosity in aqueous solution. This is because light scattering in aqueous solution is difficult to measure and shows some anomalies, through light scattering is a major method for studying solution property[19].

4.1.4. Ultracentrifugation. Ultracentrifugation is a organized technique used to find molar weight of PEs in series of 100- 10^8 gm/mol[20]. In ultracentrifugation technique, the centrifugal force present at high velocity performed on the PE solution, which comes to sedimentation of the macromolecule. sedimentation velocity can be connected to molar mass of the polyelectrolyte[21].

4.2 Electrochemical characterization

4.2.1. Potentiometricmethod. This method is utilized to guess the ion activities of H+ ion via selective and reference electrodes[22]. pKa, degree of ionization, pH are the basic capacities to characterize PEs[23].so the pKa value of polyelectrolyte has been determined experimentally by potentiometric titration system and computed with molecular dynamics and constant pH method.

4.2.2. NMR spectroscopy. To calculate the chemical arrangement of the macromolecule, NMR spectroscopy is used. NMR spectroscopy technique gives us additional data about molecular dynamics present in the bulk state[24]. NMR (Proton nuclear magnetic resonance)spectroscopy technique is used for study of the changes in the chain structure and dynamics during the temperature-induced phase separation in polymer solutions[25].The resolution which is given by the NMR method permits the signal of polycation and polyanion to be monitored distinctly for straight observation of the (LBL) growth[26].

5. Deposition methods

Polyelectrolyte multilayers were deposited by two techniques 1) Dip coating. 2) Spincoating. In dip coated PEM, an interfusion of polyelectrolyte layer was observed while flat and clearly separated layer were deposited by spin coating[27].The exchanged dip coating is one of the easy method suitable for use. To increase the speed of the deposition of drug, spray deposition and spin coating has considered as a noble option. Spin coated films are narrow, extra apparent and of advanced elasticity produced through alter dipping [28]. Layer by layer (LBL) assemblies have just undertaken wide study for drug delivery application. The benefit observed by (LBL)technique is that rate of the growth of the film has measured by nanometre scale [29]. Capsule sort of LBL method has specific attention due to their potential for targeted distribution in body. Film sort of assembly cannot usually transportable over body but film kind of assembly technique can be dominant approach for the covering of biomedical devices requires constant delivery of drug [30]. Film formed in a regular method can deposited on a various substrate in a robotized way using different techniques like alternated dip coating, spin coating or spray deposition. Films prepared up by using polyelectrolyte substances is easy deposition technique, which intensely makes differences with the difficulty of the fundamental mechanism. For the duration of the film deposition, various properties can be found by shifting the operational parameters with certain grouping of interacting species [31]

5.1 Layer-By-Layer (LBL) assembly

Gero Decher and Hong first experimentally found LBL assembly method. Coating collides and consequent shell development are some applications of LBL which was invented after several years[32].Highly efficient DDS is every time made through hybridisation of many constituent and assembly technique and every technique holds its own unique part in DDS. The (LBL) adsorption method offers vast opportunity in the suitable material selection similarly flexibility of the structural architecture for material, which is fully matched fabrication of the drug delivery substance demanding difficult drug design [33].In (LBL) self-assembly technique it frequently give rise to a linear growth, for film thickness. Collective number of polyelectrolyte deposition layer, it is detected that film thickness raises exponentially and these occurrence be observed for the weakly charged polyelectrolyte multilayer [29].LBL assemblies in recent times go through extensive research for drug delivery application. Capsule sort of LBL has been of specific concern due to their potential for directed delivery in body. Film sort of assembly cannot transportable throughout the body but this one is the dominant method for the covering of biomedical apparatus which requires essential persistent delivery of drug[34], [35]. The build-up of LbL multilayers focused by the electrostatic interaction among the opposite charged particles. Multiple function of LBL methods can be presented by the occurrence of different materials in the multilayers like PE, metallic oxides, Nano objects and catalytic enzymes[35].Benefit of polyelectrolyte capsules (PCs) as distribution transporters is that the capsules prepared by polymeric substance keep the medicine from degeneration on the way to the cell and defend the body from the damaging side reaction of the drug. For micro or nano capsules to be functional as remedial delivery of transportation. There is three main parts: (i) the encapsulation (ii) The targeting of the capsules in bio system, and (iii) the exact delivery of the drug from the capsules[7]. The properties of the LbL assembly can be moderated by regulating, nearly processing factors such as covering of material, ion strength, temperature, PH, washing and drying conditions.

5.2 Procedure of layer-by-layer method

LBL assembly system contains the repeated serial dipping of a charged substrate into solutions of oppositely-charged PE, and the adsorption of the opposite charged PE give rise to the development of a multilayer nanolayers on the substrate[36]. Throughout this process, opposite charges that are present on the surface leaving additional charges upon adsorption [37]. The charge on the surface exists on the outermost layer reformed between the anionic and the cationic state. Polyelectrolyte present in the solution form the electrostatic bond with functional group of containing charge. It later offers a driving force for successive build-up of multilayers in assembly (Fig1)[38].This result can be observed more or less irreversibly adsorbed polyelectrolyte multilayer of which thickness and interpretation can be manipulated with different type of adsorbing species, type of deposition technique and processing condition such as polyelectrolyte charge, ionic strength, salt concentration and adsorption time[39]. The development of LBL assembly multifilm are accomplished by multiple molecular interactions, which includes electrostatic force of attraction, covalent bond, hydrogen bond, van der Waals forces, hydrophobic interactions, charge-transfer interactions, host–guest interactions. Between these interactions, electrostatic force of attraction among the opposite charged components is the most frequently useful driving force used for the construction of nano structured PE multilayers. Attachment designs include electrostatic interaction, Diels alder reaction, Huisgen 1, 3 dipolar cyclo addition and maleimide Michael reaction. The use of non-covalent interactions is another way for loading drug molecules onto coated NPs. For example, electrostatic interactions between cationic-terminated self-assembly methods and anionic oligo nucleotides can be used to load DNA or RNA molecules on the particles.

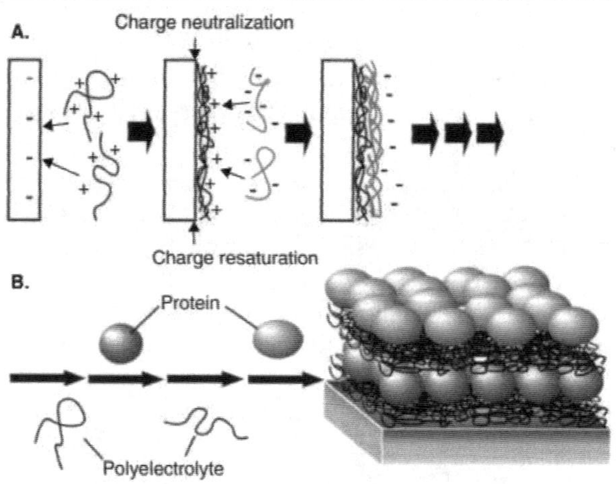

Figure 1. A LBL film of (A) polyelectrolytes and (B) polyelectrolyte/protein[32].

5.3. Chitosan based self-assembly method in DDS

Chitosan is the exceptional natural cationic polysaccharide, and it is observed that a natural polycationic linear polysaccharide obtained from the process of partial deacetylation of the naturally occurring polysaccharide, chitin. Chitosan is glucosamine (amino sugar) and N-acetyl glucosamine(amide derivative of monosaccharide glucose) units allied together by β (1 → 4) glycosidic links shown in Fig.2.

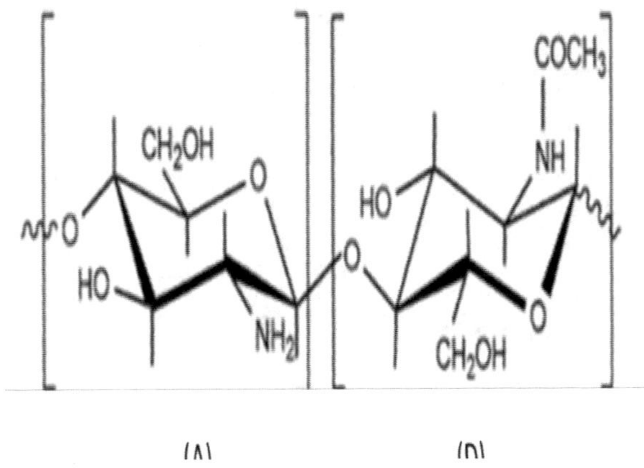

Figure 2. (A) N-acetyl glucosamine(amide derivative) unit; (D) Glucosamine unit(amino sugar)[40]

Chitosan exhibits several properties such as biocompatibility, biodegradability, non-toxic material; Chitosan also shows important biological properties such as wound healing capability, antimicrobial, and hemostatic actions[5].One of the important property of chitosan is its chemical arrangement are the hydrogen bonds (intramolecular) of CS which offer excellent resistance to heat generated due to the interaction of macromolecules[41]. ionotropic gelation, spray drying, (water-in-oil) emulsion cross-linking, formation of reverse micelle, emulsion-droplet coalescence, nanoprecipitation, and by a self-assembly mechanism such type of various method has been used to produce chitosan nanoparticle[40]. Self-assembly method is particularly useful in the preparation of injectable thermogels. Since it exhibits hydrophobic carrier system and it has exceptional cationic property, chitosan particles can interact with anionic polymers to form the PECs. Chitosan found by N-deacetylation of chitin. 2-acetomedo 2-Deoxy-β-D glucose is obtained through a β (1→4) Linkage [42]. Alkaline hydrolysis with NAOH complete for N-deacetylation of chitin at around 120°Ctemperature. 40-80% deacetylated chitosan be formed in the alkali treatment. Hence, chitosan is a hetro-polymer, which exhibit (1→4) 2-amino 2-Deoxy-β-D-glucose collected with (1→4) 2-acetamido-2-Deoxy- β-D glucose elements of the unique chitin in polymeric chain (Fig3).

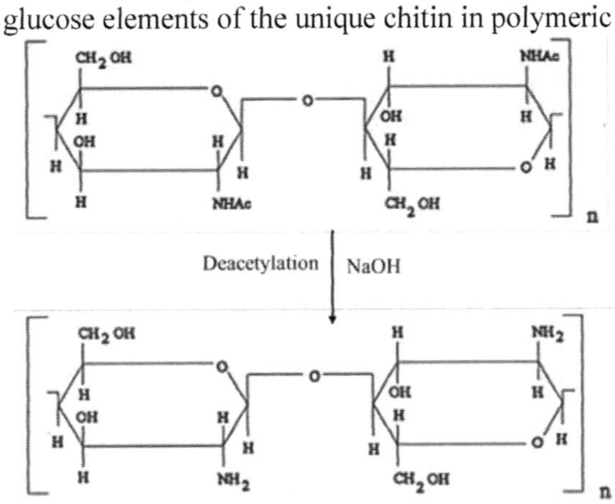

Figure 3. (Deacetylation of chitin into chitosan)[43]

Degree of deacetylation(DOD) is keyparameter. Due to DOD, solubility of polymer and solution properties observed. The DOD is the ratio of 2-amino-2-deoxy- β-D-glucose units to 2-acetamido-2-deoxy- β-D glucopyranose[44]. In this process, (chitin transformed into chitosan) the molecular weight observed to be lowered because of deacetylation (Fig4). In acidic solvent, the charge distribution changed because of the amino group formation.

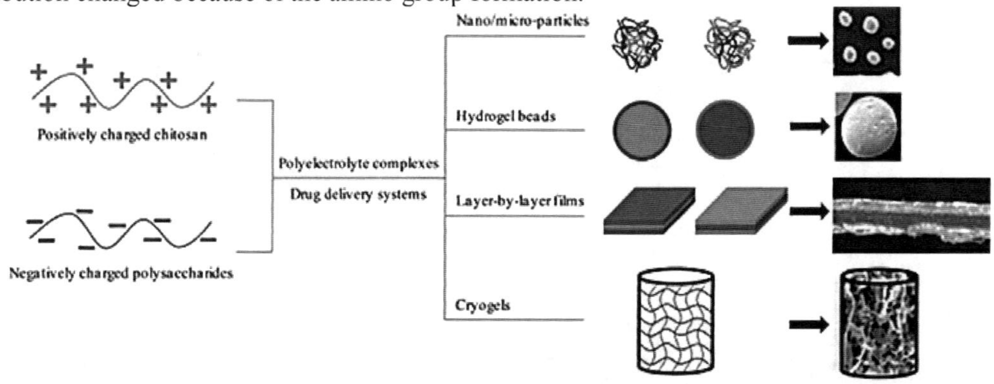

Figure 4. Development of PECs between chitosan and anionic polysaccharides [45].

6. Features of polyelectrolyte complex

Amorphous aggregates (natural or denatured proteins without forming a specific higher directive construction), held together by reversible ionic cross links with random charge reparation within the complex. Due to their surface charge, highly swollen and permeable gel particles present in aqueous solution forms stable suspensions. The ability to build up multi-layer coatings (LBL) means alternating positively and negatively charged polyelectrolytes. Polyelectrolyte family has familiar with the various important perception of PE physics like charge regulation, concomitant comparison between hydrophobicity and electrostatic contact, main forces in arrears liquid phase separation, dynamics of polyelectrolyte and transport kinetics across section[46].

7. Application of polyelectrolyte in drug delivery system

PECs have achieved great attention in several years because of their individual wide application. The improved biocompatibility of PE coated nano particles has managed to build potential biomedical application, including DDS, gene analysis, and cancer treatment[47]. Formation of coating on films and fibres, for separation and fractionation of proteins, for segregation of nucleic acid, for necessary pharmaceutical yields, as supports for catalyst and for research of microcapsules for drug delivery, These can be used as membranes[48]. PECs have important application, it contains microencapsulation of drugs, enzymes, cells and microorganism, control of proteins by the complex development and polycation complexes with poly nucleotides or oligo nucleotides as vectors in gene treatment[49]. As insoluble PECs indicate its exceptional efficiency hydrophilic soil binders, after the Chernobyl accident to subdue the developed radioactive aerosol in contaminated dusty zone[50]. The major of flocculants are their integral solid liquid separating efficiency. These varieties makes polyelectrolyte a unique class of polymers which catch extensive application[51]. The category of PE has its significant effect on the usage of particles and the contact of the particle with tissues and cells [45]. There are generally two approaches involved for drug delivery: targeted and non-targeted. In targeted delivery a lig and, usually an antibody or peptide is to be stick to the NP and target a receptor on a specific cell. Non- targeted delivery is the delivery without targeting exact cell, where the drug will be released to both healthy and diseased cell[43]. The nano constructed shells are appropriate as coating of living cells or artificial tissue, the necessities for this application are polyelectrolytes that are not poisonous to the tissue of the coated cell as well as transplantation site[52].

8. Conclusion

The existing research work in pharmaceutical nanotechnology, PECs Nano particles enhanced further developments of such system. The PENCs yet is capable considering the range of benefits they offer. Synthetic and natural PECs are rapidly emerging field of drug delivery. The biomimetic and bio motivated system have great potential to resolve any difficulty in DDS. Combining outlook from the biological field will offer a fresh regular standard for the design of polymeric DDS. The study for DDS method and the different approaches of the action signifies one of the frontline study regions. The DDS combines one or more old-style DDS with the support of engineering technologies, the system tries to produce the capability to precisely target the area where drug has delivered in the body also considered therate of release in the body at which it should need to release. If DDS depend on a biodegradable implant to carry drug medically or deep within the body, biodegradable and bio-absorbable polymers deliver a safe framework for transporting medicine without harm to the body. Polymer type DDS has well-defined as making that carries remedial matter into the body, this system recovers its efficiency viaregulating the rate of release in body, time of release, and place of release of drug in the body. DDS has reached remarkable evolution in the several years. However, remains a hard job to control drug approach into the body. The drug delivery area is developing at an exponential rate in work of research and evolution of new techniques.

9. References

[1] A. Verma and A. Verma, "Polyelectrolyte Complex- an Overview," *Int. J. Pharm. Sci. Res.*, vol. 4, no. 5, pp. 1684–1691, 2013.

[2] V. S. Rathee, A. J. Zervoudakis, H. Sidky, B. J. Sikora, and J. K. Whitmer, "Weak polyelectrolyte complexation driven by associative charging," *J. Chem. Phys.*, vol. 148, no. 11, 2018, doi: 10.1063/1.5017941.

[3] W. Mccarthy, "Polymeric Drug Delivery Techniques," *Aldrich Mater. Sci.*, no. 5, pp. 2–5, 2016.

[4] G. Tiwari *et al.*, "Drug delivery systems: An updated review," *Int. J. Pharm. Investig.*, vol. 2, no. 1, p. 2, 2012, doi: 10.4103/2230-973x.96920.

[5] A. Salama and M. El-Sakhawy, "Preparation of polyelectrolyte/calcium phosphate hybrids for drug delivery application," *Carbohydr. Polym.*, vol. 113, pp. 500–506, 2014, doi: 10.1016/j.carbpol.2014.07.022.

[6] B. P. Kumar, M. Pharm, S. Duraivel, M. Pharm, and D. Ph, "Instructions to Authors: Instructions to Authors," *Anal. Chem.*, vol. 71, no. 3, pp. 142A-143A, 1999, doi: 10.1021/ac990155t.

[7] B. S. Kim and J. W. Choi, "Polyelectrolyte multilayer microcapsules: Self-assembly and toward biomedical applications," *Biotechnol. Bioprocess Eng.*, vol. 12, no. 4, pp. 323–332, 2007, doi: 10.1007/BF02931052.

[8] S. K. Yadava, J. B. Naik, J. S. Patil, V. J. Mokale, and R. Singh, "Enhanced solubility and bioavailability of lovastatin using stabilized form of self-emulsifying drug delivery system," *Colloids Surfaces A Physicochem. Eng. Asp.*, vol. 481, pp. 63–71, 2015, doi: 10.1016/j.colsurfa.2015.04.026.

[9] L. Zhao, M. Skwarczynski, and I. Toth, "Polyelectrolyte-Based Platforms for the Delivery of Peptides and Proteins," *ACS Biomater. Sci. Eng.*, vol. 5, no. 10, pp. 4937–4950, 2019, doi: 10.1021/acsbiomaterials.9b01135.

[10] T. Liu *et al.*, "Biomedical Applications of Layer-by-Layer Self-Assembly for Cell Encapsulation: Current Status and Future Perspectives," *Adv. Healthc. Mater.*, vol. 8, no. 1, pp. 1–16, 2019, doi: 10.1002/adhm.201800939.

[11] J. Almer, *polyelectrolyte thermodynamics and rheology*, vol. 30, no. 3. 2017.

[12] T. Zhu *et al.*, "Metallo-polyelectrolytes as a class of ionic macromolecules for functional materials," *Nat. Commun.*, vol. 9, no. 1, 2018, doi: 10.1038/s41467-018-06475-9.

[13] V. S. Meka, M. K. G. Sing, M. R. Pichika, S. R. Nali, V. R. M. Kolapalli, and P. Kesharwani,

"A comprehensive review on polyelectrolyte complexes," *Drug Discov. Today*, vol. 22, no. 11, pp. 1697–1706, 2017, doi: 10.1016/j.drudis.2017.06.008.

[14] H. Rmaile, *Structure-Property Relationships in the Design, Assembly and Applications of Polyelectrolyte Multilayer Thin Films*. 2004.

[15] A. Umerska *et al.*, "Exploring the assembly process and properties of novel crosslinker-free hyaluronate-based polyelectrolyte complex nanocarriers," *Int. J. Pharm.*, vol. 436, no. 1–2, pp. 75–87, 2012, doi: 10.1016/j.ijpharm.2012.07.011.

[16] J. M. Y. Carrillo and A. V. Dobrynin, "Salt effect on osmotic pressure of polyelectrolyte solutions: Simulation study," *Polymers (Basel).*, vol. 6, no. 7, pp. 1897–1913, 2014, doi: 10.3390/polym6071897.

[17] D. Carrière, M. Dubois, M. Schönhoff, T. Zemb, and H. Möhwald, "Counter-ion activity and microstructure in polyelectrolyte complexes as determined by osmotic pressure measurements," *Phys. Chem. Chem. Phys.*, vol. 8, no. 26, pp. 3141–3146, 2006, doi: 10.1039/b602583h.

[18] E. Spruijt *et al.*, "Structure and dynamics of polyelectrolyte complex coacervates studied by scattering of neutrons, X-rays, and light," *Macromolecules*, vol. 46, no. 11, pp. 4596–4605, 2013, doi: 10.1021/ma400132s.

[19] M. Hara and A. Nakajima, "Characteristic behaviors of light scattering from polyelectrolyte in dilute solution region," *Polym. J.*, vol. 12, no. 10, pp. 701–709, 1980, doi: 10.1295/polymj.12.701.

[20] Z. Cao, S. Wu, and G. Zhang, "Dynamics of single polyelectrolyte chains in salt-free dilute solutions investigated by analytical ultracentrifugation," *Phys. Chem. Chem. Phys.*, vol. 17, no. 24, pp. 15896–15902, 2015, doi: 10.1039/c5cp02111a.

[21] Z. Luo and G. Zhang, "Sedimentation of polyelectrolyte chains in aqueous solutions," *Macromolecules*, vol. 43, no. 23, pp. 10038–10044, 2010, doi: 10.1021/ma101324j.

[22] T. Masadome, Y. Yamagishi, M. Takano, and T. Hattori, "Potentiometric titration of polyhexamethylene biguanide hydrochloride with potassium poly(vinyl sulfate) solution using a cationic surfactant-selective electrode," *Anal. Sci.*, vol. 24, no. 3, pp. 415–418, 2008, doi: 10.2116/analsci.24.415.

[23] F. Mousseau, L. Vitorazi, L. Herrmann, S. Mornet, and J. F. Berret, "Polyelectrolyte assisted charge titration spectrometry: Applications to latex and oxide nanoparticles," *J. Colloid Interface Sci.*, vol. 475, pp. 36–45, 2016, doi: 10.1016/j.jcis.2016.04.036.

[24] L. N. J. Rodriguez, S. M. De Paul, C. J. Barrett, L. Reven, and H. W. Spiess, "Fast magic-angle spinning and double-quantum 1H solid-state NMR spectroscopy of polyelectrolyte multilayers," *Adv. Mater.*, vol. 12, no. 24, pp. 1934–1938, 2000, doi: 10.1002/1521-4095(200012)12:24<1934::AID-ADMA1934>3.0.CO;2-S.

[25] R. N. Smith, L. Reven, and C. J. Barrett, "13C solid-state NMR study of polyelectrolyte multilayers," *Macromolecules*, vol. 36, no. 6, pp. 1876–1881, 2003, doi: 10.1021/ma0258408.

[26] H. Saveyn, P. M. S. Hendrickx, S. K. Dentel, J. C. Martins, and P. Van der Meeren, "Quantification of hydrolytic charge loss of DMAEA-Q-based polyelectrolytes by proton NMR spectroscopy and implications for colloid titration," *Water Res.*, vol. 42, no. 10–11, pp. 2718–2728, 2008, doi: 10.1016/j.watres.2008.01.024.

[27] F. Boulmedais, C. S. Tang, B. Keller, and J. Vörös, "Controlled electrodissolution of polyelectrolyte multilayers: A platform technology towards the surface-initiated delivery of drugs," *Adv. Funct. Mater.*, vol. 16, no. 1, pp. 63–70, 2006, doi: 10.1002/adfm.200400406.

[28] J. B. Schlenoff and S. T. Dubas, "Mechanism of polyelectrolyte multilayer growth: Charge overcompensation and distribution," *Macromolecules*, vol. 34, no. 3, pp. 592–598, 2001, doi: 10.1021/ma0003093.

[29] M. Serhan *et al.*, "Total iron measurement in human serum with a smartphone," *AIChE Annu. Meet. Conf. Proc.*, vol. 2019-Novem, 2019, doi: 10.1039/x0xx00000x.

[30] J. M. Garza *et al.*, "Multicompartment films made of alternate polyelectrolyte multilayers of

exponential and linear growth," *Langmuir*, vol. 20, no. 17, pp. 7298–7302, 2004, doi: 10.1021/la049106o.
[31] Y. Guo, W. Geng, and J. Sun, "Layer-by-layer deposition of polyelectrolyte-polyelectrolyte complexes for multilayer film fabrication," *Langmuir*, vol. 25, no. 2, pp. 1004–1010, 2009, doi: 10.1021/la803479a.
[32] K. Ariga, M. McShane, Y. M. Lvov, Q. Ji, and J. P. Hill, "Layer-by-layer assembly for drug delivery and related applications," *Expert Opin. Drug Deliv.*, vol. 8, no. 5, pp. 633–644, 2011, doi: 10.1517/17425247.2011.566268.
[33] A. Q. Alorabi *et al.*, "On-chip polyelectrolyte coating onto magnetic droplets-towards continuous flow assembly of drug delivery capsules," *Lab Chip*, vol. 17, no. 22, pp. 3785–3795, 2017, doi: 10.1039/c7lc00918f.
[34] M. M. De Villiers, D. P. Otto, S. J. Strydom, and Y. M. Lvov, "Introduction to nanocoatings produced by layer-by-layer (LbL) self-assembly," *Advanced Drug Delivery Reviews*, vol. 63, no. 9. pp. 701–715, 2011, doi: 10.1016/j.addr.2011.05.011.
[35] A. Shakiba, O. Zenasni, M. D. Marquez, and T. Randall Lee, "Advanced drug delivery via self-assembled monolayer-coated nanoparticles," *AIMS Bioeng.*, vol. 4, no. 2, pp. 275–299, 2017, doi: 10.3934/bioeng.2017.2.275.
[36] B. Li, "Advances in polyelectrolyte multilayer nanofilms as tunable drug delivery systems," *Nanotechnol. Sci. Appl.*, vol. Volume 2, pp. 21–27, 2009, doi: 10.2147/nsa.s5705.
[37] K. Ariga, Y. M. Lvov, K. Kawakami, Q. Ji, and J. P. Hill, "Layer-by-layer self-assembled shells for drug delivery," *Adv. Drug Deliv. Rev.*, vol. 63, no. 9, pp. 762–771, 2011, doi: 10.1016/j.addr.2011.03.016.
[38] M. Delcea, H. Möhwald, and A. G. Skirtach, "Stimuli-responsive LbL capsules and nanoshells for drug delivery," *Adv. Drug Deliv. Rev.*, vol. 63, no. 9, pp. 730–747, 2011, doi: 10.1016/j.addr.2011.03.010.
[39] J. K. Patra *et al.*, "Nano based drug delivery systems: Recent developments and future prospects 10 Technology 1007 Nanotechnology 03 Chemical Sciences 0306 Physical Chemistry (incl. Structural) 03 Chemical Sciences 0303 Macromolecular and Materials Chemistry 11 Medical and He," *J. Nanobiotechnology*, vol. 16, no. 1, pp. 1–33, 2018, doi: 10.1186/s12951-018-0392-8.
[40] J. P. Quiñones, H. Peniche, and C. Peniche, "Chitosan based self-assembled nanoparticles in drug delivery," *Polymers (Basel).*, vol. 10, no. 3, pp. 1–32, 2018, doi: 10.3390/polym10030235.
[41] P. Kadambi, P. Luniya, and P. Dhatrak, "Current advancements in polymer / polymer matrix composites for dental implants : A systematic review," *Mater. Today Proc.*, no. 46, pp. 740-745, 2021, doi: 10.1016/j.matpr.2020.12.396.
[42] L. Hu, Y. Sun, and Y. Wu, "Advances in chitosan-based drug delivery vehicles," *Nanoscale*, vol. 5, no. 8, pp. 3103–3111, 2013, doi: 10.1039/c3nr00338h.
[43] M. Ishihara, S. Kishimoto, S. Nakamura, Y. Sato, and H. Hattori, "Polyelectrolyte complexes of natural polymers and their biomedical applications," *Polymers (Basel).*, vol. 11, no. 4, pp. 1–12, 2019, doi: 10.3390/polym11040672.
[44] M. Marudova *et al.*, "Nanostructured polyelectrolyte multilayer drug delivery systems for buccal administration," *Bulg. Chem. Commun.*, vol. 48, no. Special Issue C, pp. 468–474, 2016.
[45] B. D. Emmanuel, N. Y. Abu-Thabit, and N. C. Ngwuluka, *Responsive polyelectrolyte complexes based on natural polysaccharides for drug delivery applications*. Elsevier Ltd., 2018.
[46] F. Le Floch, S. Belbekhouche, J. Oniszczuk, and B. Carbonnier, *Smart polyelectrolyte multilayer coatings for drug delivery*. Elsevier Inc., 2019.
[47] Z. Folchman-Wagner, J. Zaro, and W. C. Shen, "Characterization of polyelectrolyte complex formation between anionic and cationic poly(amino acids) and their potential applications in

pH-dependent drug delivery," *Molecules*, vol. 22, no. 7, 2017, doi: 10.3390/molecules22071089.
[48] B. G. De Geest, G. B. Sukhorukov, and H. Möhwald, "The pros and cons of polyelectrolyte capsules in drug delivery," *Expert Opin. Drug Deliv.*, vol. 6, no. 6, pp. 613–624, 2009, doi: 10.1517/17425240902980162.
[49] V. Rana, P. Rai, A. K. Tiwary, R. S. Singh, J. F. Kennedy, and C. J. Knill, "Modified gums: Approaches and applications in drug delivery," *Carbohydr. Polym.*, vol. 83, no. 3, pp. 1031–1047, 2011, doi: 10.1016/j.carbpol.2010.09.010.
[50] A. P. R. Johnston, C. Cortez, A. S. Angelatos, and F. Caruso, "Layer-by-layer engineered capsules and their applications," *Curr. Opin. Colloid Interface Sci.*, vol. 11, no. 4, pp. 203–209, 2006, doi: 10.1016/j.cocis.2006.05.001.
[51] M. Müller, B. Keßler, J. Fröhlich, S. Poeschla, and B. Torger, "Polyelectrolyte complex nanoparticles of poly(ethyleneimine) and poly(acrylic acid): Preparation and applications," *Polymers (Basel).*, vol. 3, no. 2, pp. 762–778, 2011, doi: 10.3390/polym3020762.
[52] S. Raj, P. Kumar Sharma, and R. Malviya, "Pharmaceutical and Tissue Engineering Applications of Polyelectrolyte Complexes," *Curr. Smart Mater.*, vol. 3, no. 1, pp. 21–31, 2018, doi: 10.2174/2405465803666180409130241.

IOP Conf. Series: Materials Science and Engineering 1183 (2021) 012008 doi:10.1088/1757-899X/1183/1/012008

Effect of Zirconium Dibromide and Aluminum Oxide Hybrid Particle Reinforcement on the Mechanical, Microstructure and Wear Properties of AA6063 Surface Metal Matrix Composite

R. Raja[1], Sabitha Jannet[1*], Christo joy[2], Jeffin Johnson[3]

[1]Assistant Professor, Mechanical Department, Karunya Institute of Technology and Sciences, Coimbatore-641114, Tamilnadu, India.
[2]UG Scholar, Mechanical Department, Karunya Institute of Technology and Sciences, Coimbatore-641114, Tamilnadu, India.
[3]Assistant Professor, Mechanical Engineering Department, Rajagiri School of Engineering & Technology, Kochi, Kerala, India.

[*]Corresponding author Email: sabithajannet@gmail.com

Abstract. In the present study Aluminum 6063 as the matrix and with Zirconium dibromide and Aluminum oxide as the reinforcement particle Metal Matrix surface composites were fabricated by friction stir processing. By the addition of reinforcements such as Zirconium Dibromide and Aluminum Oxide used as the strengthening agents during FSP process certainly made the MMC more wear resistant and harder than the base matrix. The effect of varying reinforcement percentage on the Mechanical wear and microstructure properties were studied. The Micro hardness showed an increasing trend in the nugget zone. The wear rate decrease with increase in volume percentage of hybrid reinforcement. FESEM images of the samples displayed uniform distribution of the particle in the matrix.And also reiterated the formation of hybrid particles.

Keywords: *Friction stir processing, Hybrid surface composite, Wear, Microhardness, FESEM*

1. Introduction

Aluminum alloys are the answer for the culmination of searches made by material scientist for the cost effective functional material. Aluminum Metal Matrix has been replacing the conventional engineering materials because of its properties which could match the anticipations of the industrial requirements. Friction stir processing is a derivative of FSW and has been extensively tried to fabricate surface composites. The proper dispersion of the strengthening agents while processing the metal matrix composites is very much required since it will ensure not only its upgradation over various characteristics like improved wear resistance and hardness but also over its metallurgical attributes. Q. Liu et al[1] reported that with increase in the addition of MWCNTs there was gradual increase in the tensile strength and microhardness. H. I. Kurt et al[2] inferred the effect of various proportion of hybrid particles and the process parameters such tool traverse speed, rotational speed and tool profile's effect on the properties. I Dinaharan, et al[3] analyzed the mechanical and wear properties of aluminum copper cast . H. I. Kurt et al[4] studied the wear behaviour of AA6061 T6

Content from this work may be used under the terms of the Creative Commons Attribution 3.0 licence. Any further distribution of this work must maintain attribution to the author(s) and the title of the work, journal citation and DOI.
Published under licence by IOP Publishing Ltd

reinforced with Graphite and Titanium Carbide microparticles. S. J. Abraham et al[5] reported wear mechanical behaviour and thermo mechanical studies on the AA6063 reinforced with Quartz particles. M. Balakrishnan et al[6] studied the tensile behaviour and microstructure analysis of Aluminum ferrite reinforced AA 6061.

The micro features of metals can be enhanced through friction stir processing method. The processing factors, tool's shape and size and associated joint structure make use of the prominent effect on heat dissipation and the dispersion of material, consequently enhancing the interior of the metal matrix body even from the atomic level. For that purpose a fast moving tool which is rotated and forced against the metal matrix along with the reinforcements so that the region which is exposed to the vigorously acted tool would undergo plastic deformation uniformly under the frictional heat. Properties of materials like aluminum, copper, magnesium have been upgraded by the process of FSP extensively. The groove size and the fineness of the strengthening agents affect and even could predict the final characteristic of the composite. R Beygi et al[11] found out that the fineness of the titanium carbide powder increased drastically the yield strength of the aluminum composite. Size of the reinforcements largely aided the formation of a aluminum composite with superior strength than the base matrix.

In this present study Aluminum alloy 6063 is reinforced with Zirconium dibromide and aluminum oxide micro particles and material and mechanical characterization were conducted.

2. Materials and Methods

Hybrid reinforcement was achieved by a Magnetic stirrer. Aluminum oxide with zirconium dibromide was used in equal percentage. Acetone was used for the mixing in a magnetic stirrer which was stirred 1560 rpm to obtain hybrid particles. V shaped grooves were machined using EDM of 0.3mm,0.6mm,0.9mm widths and 5mm depth and the particles were packed into them. A vertical milling machine was used to carry out the FSP at Vigshan tools, Coimbatore. The samples thus prepared underwent a single run of FSP with a pinless tool. The pinless tool solves the purpose of the powder being intact in the groove. Subsequently 3 runs of FSP were done on the samples with a cylindrical tool of shoulder diameter 18 mm and pin diameter of 6 mm with a pin length of 5 mm. The FSPed samples were then cut in dimensions of 30 mm X 30 mm for wear and microhardness tests. The wear tests were conducted on Pin on Disc apparatus. The microhardness tests were also conducted using Vickers hardness testing machine. The samples were then polished and etched to obtain detailed microstructure from FESEM images. The various compositions of the reinforcement and matrix is shown in detail in Table 2.

Table 1. Chemical composition of Al6063

Elements	Si	Fe	Cu	Mn	Mg	Cr	Zn	Ti	Al	others
Weight %	0.2 to 0.6	0.35	0.10	0.10	0.45 to 0.9	0.10 max	0.10	0.10	98.75	0.15

Table 2. Composition details of the composites

Sl No	Sample	Composition	
		AA6063(%)	Zirconium Dibromide + Aluminum Oxide(%)
1	A1	95	5
2	A2	90	10
3	A3	85	15

3. Results and Discussion

3.1. Microhardness

The three samples which were prepared with 5%, 10% and 15% of $ZrBr_2$+Al_2O_3 had undergone microhardness testing in order to analyze whether any enhancement has obtained in the regions nearby and within the stir zone.

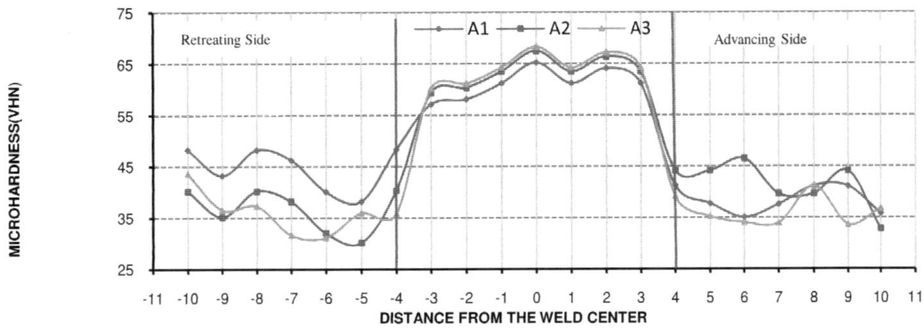

Figure 1. Micro hardness variation on the three zones of the hybrid surface composite

The micro hardness test indicated that the samples showed an increase in the micro hardness in the stir zone. From the figure 1 it can be inferred that the increase in the micro hardness in the stir zone compared to the advancing and retreating side is majorly due to the grain size reduction. Cao R. et al7, this would have occurred due to reinforcement of hybrid particles and the severe plastic deformation which the samples would have undergone during FSP process. In the FSP zone hardness increases due to high heat produced in the zone and because of high heat, dispersion of reinforcement has taken place in a homogeneous manner and thereby no chance for porosity in that particular region it enables the FSP zone to have high hardness compared to the other zone.

A considerable variation in the hardness of the composite and the base matrix was observed, in addition, the micro hardness of material away from the stir zone was found out to be 27 VHN whereas that within the stir zone is as high as 66 VHN. By FSP due to the controlled rotation of the processing tool had ensured grain refinement and thereby led to an increase in the micro hardness of the composite. There was a considerable mix up of the reinforcements to strengthen up the composite during the dispersion and led to the uniform distribution of the elements within the nugget zone.

3.2. Wear Rate

Using the pin on disc apparatus, the wear rate of the three samples were measured, in addition, it was observed that the wear rate gradually decreased with increase in % of $ZrBr_2$+Al_2O_3 and finally the sample with 15% showed the least wear rate among the taken samples.

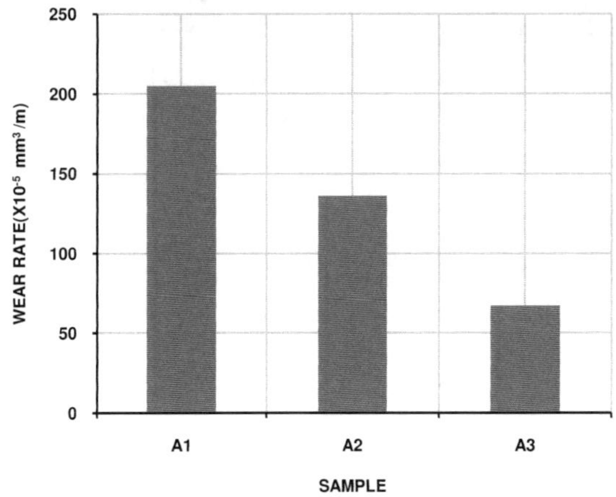

Figure 2. Wear Rate

The appreciable interfacial bonding between the reinforcement and the matrix in advents the sliding of the surface. The contact surface between the pin and the aluminum alloy 6063 is reduced due to the presence of hybrid particles can be one of the reasons for reduced wear rate also reported by Maurya S et al[8] and Vijayan et al[9], this can also be attributed to the reason that hard materials usually have lesser wear rate compared to soft materials.

Usually, aluminum being a softer material tends to wear out more than other materials in use even though it has a better possibility in aircraft and automobile applications if its wear rate could be controlled by the addition of proper reinforcements. Consequently, the particles will get into the aluminum matrix especially in areas where it would wear out more easily thereby enhancing the resistance to wear against the contact.

3.3. Microstructure Analysis

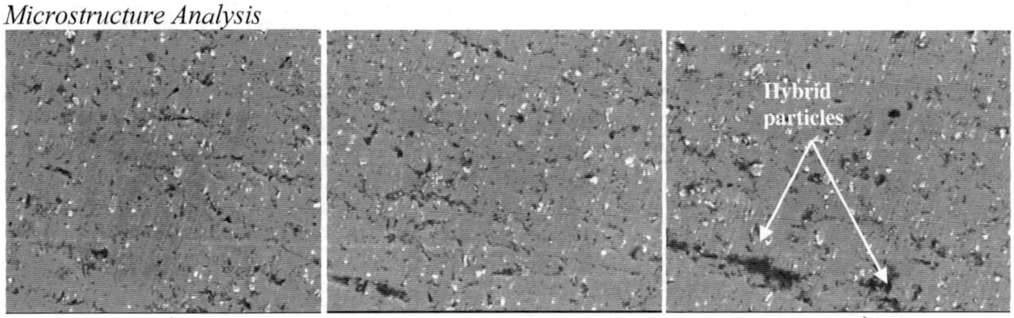

a) b) c)

Figure 3. FESEM images of the microstructure of composite a)A1 b)A2 c)A3

The microstructure from the FESEM images shown in Figure 3 (a-c) indicates the distribution of the reinforcement particle at various volume percentages. The images displayed show an uniform distribution of the particles also analyzed by A Sert et al[10].The matrix underwent severe plastic deformation leading to the stirring and uniform distribution of the particles. It was also observed a few cracks in the images which might be due to non-uniform cooling after the process.

The severe plastic distortion by the rotating tool had brought the reinforcements into the matrix especially in the stir zone where dynamic recrystallization had taken place during the FSP process. When the temperature rose above recrystallization temperature because of the friction between tool and substrate made the material more plastic rather than liquid which led to uniform as well as proper dispersion of the strengthening agents in to the aluminum matrix.

IOP Conf. Series: Materials Science and Engineering 1183 (2021) 012008 doi:10.1088/1757-899X/1183/1/012008

4. CONCLUSION

Zirconium dibromide along with Aluminum oxide micro particles were used to reinforce Aluminum alloy 6063.Friction stir processing was implemented to fabricate the Aluminum Metal Matrix surface composites.

1. The wear test for the composites were conducted and the result shows that wear rate decreases as the volume percentage of the reinforcement added into the composite increases, what's more, it was found out that Al6063 with 15% of $ZrBr_2$+Al_2O_3 reported least wear rate among the samples due to the considerable presence of hybrid particles within the fabricated composite.

2. The result for the micro hardness test clearly shows that hardness value increases as the volume percentage of the reinforcement added to the composites increases, in fact the grain refinement which had taken place because of the rotation of the tool against the solid substrate with the strengthening elements supported the case for such an improvement in the mechanical characteristics. The microhardness of the sample with 15% of $ZrBr_2$+Al_2O_3 was observed to be the largest among the samples due to the fact that added hybrid particles which dispersed uniformly within the matrix had kept the entire fabricated composite intact without any further dislocation towards the slipping plane. The microhardness was raised to above 65 VHN from 25 VHN in the nugget zone and it is considered to be an appreciable amount of enhancement happened to the property after FSP.

3. The FESEM analysis indicates the presence and uniform distribution of reinforcement particles in the matrix metal. Being a solid state process, due to the intense plastic deformation under frictional heat led to the dynamic recrystallization of the material and thereby distributes the hybrid micro particles uniformly especially for the sample with 15% of $ZrBr_2$+Al_2O_3 which was observed from the FESEM images.

References

[1] Q. Liu, L. Ke, F. Liu, C. Huang & L. Xing, Microstructure and mechanical property of multi-walled carbon nanotubes reinforced aluminum matrix composites fabricated by friction stir processing, Mater. Des 45 (2013) 343–348.

[2] H. I. Kurt, Influence of hybrid ratio and friction stir processing parameters on ultimate tensile strength of 5083 aluminum matrix hybrid composites, Compos. Part B Eng., 93(2016) 26–34.

[3] I Dinaharan, M. Balakrishnan, J. David Raja Selvam& E. T. Akinlabi, Microstructural characterization and tensile behavior of friction stir processed AA6061/Al2Cu cast aluminum matrix composites, J. Alloys Compd. 78(2019) 270–279.

[4] H. I. Kurt, m. Oduncuoglu& r. Asmatulu, Wear Behavior of Aluminum Matrix Hybrid Composites Fabricated through Friction Stir Welding Process, J. Iron Steel Res. Int. 23(2016) 1119–1126.

[5] S J Abraham, S C R. Madane, I. Dinaharan& L. J. Baruch, Development of quartz particulate reinforced AA6063 aluminum matrix composites via friction stir processing, J. Asian Ceram. Soc. 4, (2016) 381–389.

[6] M. Balakrishnan, I. Dinaharan, R. Palanivel & R. Sathiskumar, Effect of friction stir processing on microstructure and tensile behavior of AA6061/Al 3 Fe cast aluminum matrix composites, J. Alloys Compd. 38(2018) 531–541.

[7] X. Cao, Q. Shi, D. Liu, Z. Feng, Q. Liu & G. Chen, Fabrication of in situ carbon fiber/aluminum composites via friction stir processing: Evaluation of microstructural, mechanical and tribological behaviors, Compos. Part B Eng .139(2018) 97–105.

[8] R. Maurya, B. Kumar, S. Ariharan, J. Ramkumar & K. Balani, Effect of carbonaceous reinforcements on the mechanical and tribological properties of friction stir processed Al6061 alloy, Mater. Des. 98(2016)155–166.

[9] S Vijayan, J P Lalith Gnanavel, G Selvakumar & SRK Rao, Study on surface characteristics of Friction Stir processed AZ91 with Titanium Carbide micro particles, IJEMS. 26 June-August (2019) 205-210.

[10]A Sert, O N Celik, Wear Behaviour of SiC reinforced surface composite Al 7075-T651Aluminum Alloy produced using Friction stir processing, IJEMS. 21 (2014) 35-43.

[11]R. Beygi, M. Zarezadeh Mehrizi, and G. Eisaabadi B, Friction Stir Processing of Al with Mechanically Alloyed Al-TiO2-Graphite Powder: Microstructure and Mechanical Properties, Journal of Materials Engineering and Performance, February 2017, 10.1007/s11665-017-2552-7.

AUTHOR INDEX